Vehicle Electronic Systems and Fault Diagnosis

Vehicle Electronic Systems and Fault Diagnosis

A practical guide for vehicle technicians

Allan W. M. Bonnick

MPhil, CEng, MIMechE, FIMI, MIRTE

Routledge
Taylor & Francis Group

LONDON AND NEW YORK

First published by Butterworth-Heinemann

This edition published 2011 by Routledge
2 Park Square, Milton Park, Abingdon, Oxon OX14 4RN
711 Third Avenue, New York, NY 10017, USA

Routledge is an imprint of the Taylor & Francis Group, an informa business

First edition by Arnold 1988

British Library Cataloguing in Publication Data
A catalogue record for this book is available from the British Library

Library of Congress Cataloging-in-Publication Data
A catalog record for this book is available from the Library of Congress

ISBN: 978-0-415-50301-3

Contents

Preface

The purpose of this book is to provide an understanding of electronics at a level that is appropriate to the needs of vehicle technicians who are likely to be involved in the maintenance and repair of vehicles equipped with electronically controlled systems. This means just about every vehicle technician, because electronics has featured as a significant part of vehicle technology for many years and the electronics content of vehicles is set to increase for the foreseeable future.

Used vehicles which are long out of warranty need maintenance and repair and vehicle technicians in many different work situations will be faced with electronic fault diagnosis and repair because owners may wish to have their vehicle repaired at their local independent garage rather than having to take it to the nearest franchised garage.

Most of the electronics which are to be found on modern vehicles are there to act in some controlling function as part of a system, and the vehicle itself is an assemblage of systems. This means that virtually all vehicle electronic systems have a common basic structure. That is to say, they consist of sensors, actuators, interconnecting circuits and a control unit. An understanding of this idea is helpful when attempting to select a strategy for faultfinding and maintenance.

Even at this stage (1998) electronics forms a minority part of a vehicle, and predictions (as shown in Figure P.1) suggest that the proportion of a vehicle that is electronic is not likely to grow disproportionately in the remainder of this century.

Experience indicates that vehicle technicians can perform accurate diagnosis and repair of vehicle electronic systems when they have only fairly basic knowledge of electronics; this book aims to provide most of that basic knowledge.

The descriptions in this book attempt to avoid the mathematics associated with the study of electronics. Ideally, some of the principles that are referred to, such as switching, and amplification properties of transistors, would be examined by setting up circuits and measuring the effects of certain inputs, etc., on the outputs of the circuit. Those readers who wish to study electronics in greater depth are advised to take a course of study of practical electronics. Those who are unable to attend a course might wish to consider studying at home: publications that will be

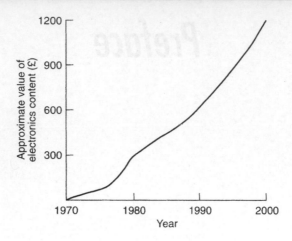

Fig. P.1   Probable growth trend in amount of electronics on a motor vehicle

of assistance in this respect are: *Automotive Electrical Maintenance*, by Stewart Robinson; *Automobile Electrical & Electronic Systems*, by Tom Denton; and *Elementary Electronics*, by Mel Sladdin and Alan Johnson, all published by Hodder Headline (Arnold).

Vehicle manufacturers offer comprehensive training on their electronic products, and it is essential that use is made of such facilities at every opportunity. It is hoped that this book will assist in improving the basic knowledge of technicians who wish to further their knowledge prior to taking advantage of manufacturers' courses.

Allan Bonnick

Acknowledgements

The illustrations in the following lists are the copyright of the respective organisations and are reproduced with their permission:

Rover Group: Figures 1.1, 2.5(a), 2.8, 2.9, 2.10, 2.18, 2.19, 3.5(d), 3.6, 3.9, 3.10, 3.17, 3.19, 3.20, 3.21, 3.22, 3.24, 3.25, 3.26, 3.27, 3.28, 3.29, 4.14, 4.15, 4.20, 4.27, 4.28, 4.29, 4.30, 4.31, 4.32, 4.33, 4.34, 4.35, 4.36, 4.37, 4.38, 4.40, 6.23, 6.27, 6.28, 6.30, 6.31, 6.33, 6.34, 6.41, 6.45, 6.46, 8.3, 8.5, 8.12, 8.14, 8.15, 8.16, 8.27, 8.46

WABCO: Figures 5.7, 6.9, 6.10, 7.13, 7.14, 7.15

Other illustrations, as listed below, were supplied by:

Avometer: Figure 3.18

Crypton: Figures 6.44, 7.3

Johnson Matthey: Figures 4.4, 4.5

LucasVarity (formerly Lucas): Figures 2.2, 2.11, 2.12, 3.2, 3.3, 3.4, 3.13, 3.23, 4.10, 4.11, 4.12, 4.13, 4.17, 4.19, 4.22, 4.23, 4.26, 8.20, 8.47

Toyota: Figures 1.3, 1.5, 2.13, 2.14, 4.3, 4.6, 4.7, 4.24, 4.25, 5.2, 5.3, 6.1, 6.2, 6.3, 6.4, 6.5, 6.6, 6.7, 6.8, 6.29, 6.32, 6.35, 6.36, 6.37, 6.38, 6.47, 6.48, 8.9, 8.17, 8.39, 8.40, 8.44(b), 8.45(a), 8.45(b), 8.45(c), 9.7

I also wish to thank all of the friends, family and associates who have provided information and support in the preparation of this book. The associates who permitted me to take photographs while they were performing diagnosis are acknowledged in the text.

Acknowledgements

The illustrations in the following lists are the copyright of the respective organisations and are reproduced with their permission

Other illustrations, as listed below, were supplied by:

1
Structure of the vehicle system

Before getting involved with electronics, it is helpful to remember that much of a motor vehicle remains fundamentally unchanged. For example, engines have pistons and poppet valves, brakes have drums or discs, suspension systems have springs, of some sort, and gearboxes have spur gears or epi-cyclic gear trains – as did the model T Ford.

Traditional and electronic skills required

Much of the work that technicians perform is therefore traditional, i.e. checking brake pads and discs for wear, routine servicing such as oil and filter changes, etc. Failure of a petrol engine to start on a wet morning is just as likely to be due to damp leads as ever it was. A persistent misfire may well be due to a spark plug failure, or loss of compression arising from a burnt valve, or some other mechanical failure. Just because a vehicle is fitted with a number of electronic devices does not mean that one should immediately suspect them, or assume that because they are there nothing can be done to remedy a problem.

Figure 1.1 shows part of a list of service tasks for a fairly modern vehicle; it also shows the 'traditional' nature of the work that has to be done. If a technician does this work properly, i.e. safely, methodically and thoroughly, with proper checks being performed at appropriate stages, then he/she already possesses some of the important attributes needed for effective work on electronically controlled systems.

It is also important to remember that there are well over 20 million vehicles on the road in Britain, and that the average life of a vehicle is estimated at 10 years. Older vehicles generally require more repairs than newer ones, which means that there will probably be quite a bit of repair work to be done in the future. Much of this repair work is likely to be on systems that have an electronics element and it is important for vehicle repairers to be able to deal with this aspect, as well as the 'traditional' aspects. This means that a mechanic needs to have good all-round 'traditional' skills backed up by the skills necessary to cope with the 'electronic' aspect. Throughout this text, the intention is to show that the 'average' well-

ROVER

Customer Name		Date	
Invoice/Job No	Vehicle Reg No	Speedometer Reading	

After Sales Service 1,000 miles/ 1 500 km
Lubrication Service At 6 months or 6,000 miles/10000 km then every 12 months or 12,000 miles/ 20000 km thereafter
Main Service At 12 months or 12,000 miles/20000 km then every 12 months or 12,000 miles/ 20000 km thereafter

IMPORTANT
* The speedometer reading and age of the car must be checked before commencing the service and those items requiring attention at differing intervals must be carried out at the mileage or period specified in the instructions.
* Under severe operating conditions certain items such as brake fluid may need to be serviced more frequently than the intervals shown below. Refer to Driver's Handbook or consult your Dealer.
* Any additional work found necessary as the result of a check is subject to extra cost.
† Starting at the right-hand front wheel, complete these operations at each wheel.

After Sales Service
| Lubrication Service
| | Main Service

Correct ✓
Incorrect ✗

01 Fit car protection kit
02 Check condition & security of seats & seat belts
03 Check operation of seat reclining, tilt & latching mechanisms
04 Drive on lift; stop engine
05 Check operation of all lamps
06 Check operation of horn(s)
07 Check operation of warning indicators
08 Check/adjust operation of screen/headlamp washers
09 Check operation of screen wipers & condition of blades
10 Check security/operation of foot & hand brakes; release fully after checking
11 Open bonnet; fit wing covers
12 Raise lift to convenient working height, with wheels free to rotate
13 Remove hub caps†
14 Mark stud to wheel relationship†
15 Remove wheels & inspect for damage†
16 Check tyre tread depth NSF [ ] mm OSF [ ] mm OSR [ ] mm NSR [ ] mm†
17 Check tyres for uneven wear†
18 Check tyres visually for external cuts in fabric, exposure of ply or cord structure, lumps or bulges†
19 Check/adjust tyre pressures†
20 Inspect brake pads for wear. Renew pads if necessary (front & rear) if pads are removed, check condition of discs & calipers†
21 Check/adjust handbrake
22 Grease wheel mounting spigots & refit wheels in original position†
23 Check tightness of wheel fastenings & refit hub caps†
24 Raise lift to convenient working height
25 Drain engine oil
26 Drain manual/automatic transmission oil every 2 years/24,000 miles/40000 km
27 Check visually brake & clutch hoses, pipes & unions for chafing, cracks, leaks & corrosion. Check pipes are correctly clipped
28 Visually check fuel tank, fuel lines & connections
29 Check security & condition of suspension joints & fixings
30 Check security & condition of drive shaft gaiters
31 Check security & condition of steering column, rack, joints & gaiters
32 Visually check underbody sealer for damage. Pay particular attention to areas of apparent repair
33 Check for fluid leaks from dampers
34 Renew oil filter element
35 Check tension (using gauge) & condition of alternator drive belt
36 Renew manual transmission oil every 2 years/24,000 miles/40000 km
37 Check/top up manual transmission oil

Fig. 1.1 A section of a job sheet

motivated conventional mechanic is quite capable of acquiring the ability to deal with the electronic element of vehicle repair and maintenance.

It is tempting to believe that the problem of electronics can be 'got round' by employing an electronics specialist. This might just about be feasible in a very large workshop, but for the majority of garages it is unlikely to be commercially viable. So it falls to the vehicle technician to extend her/his range of skills to encompass electronics and thus be competent to tackle the full range of service and repair operations required by motorists and vehicle operators. Knowledge of electronics alone, in isolation from the other skills that a vehicle technician needs in order to perform routine work, is not sufficient because very often the failure of an electronically controlled system is not due to the electronics but to some aspect of the conventional vehicle technology, e.g. anti-lock brakes playing up because of a wheel-bearing failure.

It is important to keep this in mind because the competent vehicle technician needs a platform of traditional skills to perform the bulk of his work. In order to cope with modern vehicle systems, however, these traditional skills need enlarging to encompass the electronics that are to be found on current and future vehicles.

Electronic systems contain elements in common and, in the first instance, I think it useful to review some of the systems currently in use on vehicles as this will serve to show what these 'elements in common' are.

Petrol injection systems

When dealing with vehicle technology it is common practice to start with the engine; so for this review of electronic systems I shall start with a petrol injection system. The petrol injection system shown in Figure 1.2 is of the multi-point type: that is to say, there is a separate injector for each cylinder of the engine. Petrol injection is widely used, in conjunction with an exhaust catalyst, to provide effective control of exhaust emissions.

At this stage I want to draw your attention to some specific parts of Figure 1.2. Note the inlet valve, above the piston. Just above the inlet valve, and to the right, is a fuel injector. This injector is operated by electro-magnetism in accordance with electrical pulses which are transmitted by the ECU (electronic control unit). The black line with the arrow on it is the cable (wire) that carries the electric current. The injector could be described as an actuator. To the left of the piston is the water jacket and there you will see the coolant temperature sensor. Here the black line, with the arrow pointing to the electronic control unit (ECU), carries the electrical signal that represents temperature to the ECU. The air-flow meter is another sensor.

The ECU is central to the operation of the system because it is here that the computing capacity is held, which enables the system to function.

This brief examination of the fuel injection system shows that it has four basic elements:

- actuators (injectors)
- sensors
- cables (connecting the elements together)
- electronic control unit

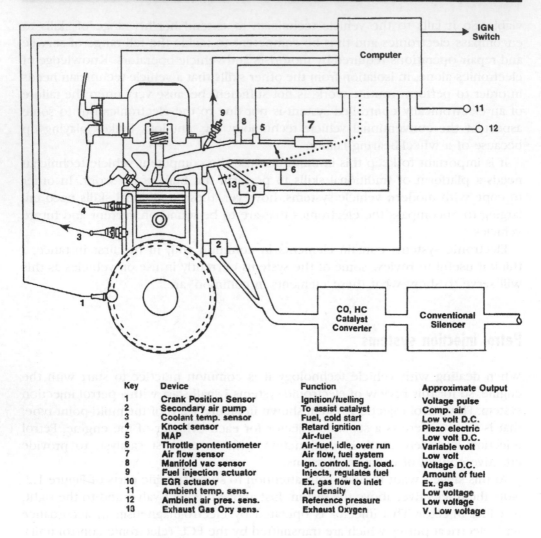

Key	Device	Function	Approximate Output
1	Crank Position Sensor	Ignition/fuelling	Voltage pulse
2	Secondary air pump	To assist catalyst	Comp. air
3	Coolant temp. sensor	Fuel, cold start	Low volt D.C.
4	Knock sensor	Retard ignition	Piezo electric
5	MAP	Air-fuel	Low volt D.C.
6	Throttle pontentiometer	Air-fuel, idle, over run	Variable volt
7	Air flow sensor	Air flow, fuel system	Low volt
8	Manifold vac sensor	Ign. control. Eng. load.	Voltage D.C.
9	Fuel injection actuator	Injects, regulates fuel	Amount of fuel
10	EGR actuator	Ex. gas flow to inlet	Ex. gas
11	Ambient temp. sens.	Air density	Low voltage
12	Ambient air pres. sens.	Reference pressure	Low voltage
13	Exhaust Gas Oxy sens.	Exhaust Oxygen	V. Low voltage

Fig. 1.2 An engine management system

Anti-lock brakes

Now let us take a brief look at another commonly used system – anti-lock brakes (ABS). ABS is used to provide enhanced braking in difficult driving conditions. Figure 1.3 shows a three-dimensional diagram of a vehicle equipped with ABS.

Here again, I want you just to pick out the main components: we have wheel-speed sensors, the hydraulic pump and accumulator (actuator), and the ABS computer (ECU), and all of these components are connected together by cable. Thus, this system has the same four basic elements as the fuel injection system, i.e.:

- actuator
- sensors

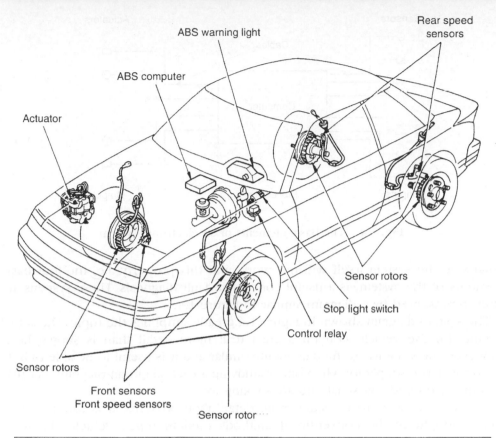

ABS warning light

Rear speed
sensors

ABS computer

Actuator

Sensor rotors

Stop light switch

Control relay

Sensor rotors

Front sensors
Front speed sensors

Sensor rotor

Component	Function
Front speed sensors	Detect wheel speed of each of the left and right front wheels.
Rear speed sensors	Detect wheel speed of each of the left and right rear wheels.
Stop light switch	Detects the brake signal and sends it to the ABS computer.
ANTI-LOCK warning light	Lights up to alert the driver when a malfunction has occured in the Anti-lock Brake System.
Actuator	Controls the brake fluid pressure to each disc brake cylinder by signals from the ABS computer.
ABS computer	It calculates acceleration and deceleration by the signals from each speed sensor and sends signals to the actuator to control brake fluid pressure.

Fig. 1.3   Anti-lock braking (Toyota)

- ECU (computer)
- interconnecting cables

It seems that the majority of vehicle electronic systems have a similar structure: it is useful to know this because each of these four elements must work properly

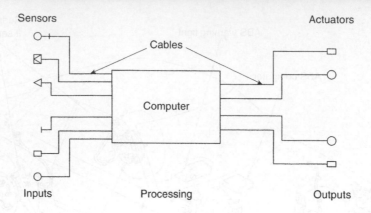

Fig. 1.4 The basic elements of an electronic system

otherwise the system itself will not work, and this property, i.e. the four basic elements of the system, is a useful concept in fault diagnosis. These systems are often represented by a diagram similar to Figure 1.4.

The system diagram shown in Figure 1.4 greatly simplifies the topic. The actual circuits on the vehicle contain a great deal more detail than is shown here. However, the systems are fundamentally similar and it is useful to be able to hold on to this basic simplicity when one is studying an actual circuit diagram, because it helps to remind you what you are looking for.

Before moving on to consider more detail about systems, it is useful to give some thought to the conventional methods used to repair vehicle electronic systems.

Electronic system repair

Individual components such as the controller (ECU), some sensors and actuators are not designed to be repaired in garages. In most garages the function of the technician is to determine which component of an electronic system is defective and to replace that component correctly. The actual method for testing a system is dependent on the diagnostic system that relates to a particular vehicle. For example, it may be that the vehicle has some built-in diagnostic capacity (on-board diagnostics, OBD) or it may rely on off-board diagnostics, or a mix of both. Whatever the case, these diagnostic systems will generally point the operator to an area of a system in which a defect lies, e.g. 'coolant sensor circuit fail'. This does not necessarily mean that the coolant sensor is defective. It could be, but it could also mean that any part of the circuit between the sensor and the test point has a defect in it. This is one reason why it is very important to know how the elements of a system are dependent on other parts of the system. Figure 1.5 indicates how, with power switched off, the circuit between the sensor and the ECU may be tested. Similar checks can be performed on the outlet (actuator) side of the ECU.

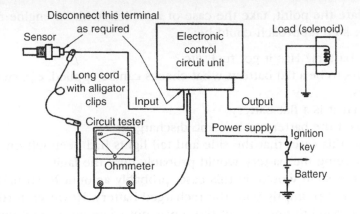

Fig. 1.5 Testing part of a circuit using a multi-meter (Toyota)

Before moving on to consider vehicle electronics in more detail it is wise to remember that much of the skill required to perform diagnosis and repair of electronic systems is the same as that which is required for good quality work of any type. By this I mean it is important to be methodical; it is unwise to start testing things randomly, or even to try changing parts in the hope that you might hit on the right thing by chance.

Many people do work methodically and they probably employ a method similar to the 'six step' approach, which is a good, common-sense approach to problem solving in general. The 'six step' approach provides a good starting reference, although it requires some refinement when used for diagnosis of vehicle systems.

We will briefly consider the 'six steps' and at a later stage take into account the refinements that are considered necessary for vehicle systems.

The 'six step' approach

This 'six step' approach may be recognised as an organised approach to problem solving in general. As quoted here it may be seen that certain steps are recursive. That is to say, it may be necessary to refer back to previous steps as one proceeds to a solution. Nevertheless, it does provide a proven method of ensuring that vital steps are not omitted in the fault tracing and rectification process. The six steps are:

1. Collect evidence.
2. Analyse evidence.
3. Locate the fault.
4. Find the cause of the fault and remedy it.
5. Rectify the fault (if different from 4).
6. Test the system to verify that repair is correct.

Just to illustrate the point, take the case of a vehicle with an engine that fails to start. The 'six step' approach could be:

1. Is it a flat battery? Has it got fuel, etc.?
2. If it appears to be a flat battery, what checks can be applied, e.g. switch on the headlamps.
3. Assume that it is a flat battery.
4. What caused the battery to become discharged?
5. Assume, in this case, that the side and tail lights had been left on. So, in this case, recharging the battery would probably cure the fault.
6. Testing the system would, in this case, probably amount to ensuring that the vehicle started promptly with the recharged battery. However, further checks might be applied to ensure that there was not some permanent current drain from the battery.

I hope that you will agree that these are good, common-sense steps to take and I am sure that most readers will recognise that these steps bear some resemblance to their own method of working.

I shall refer to these steps again. But for the time being I wish to return to the idea of the common basic structure of vehicle electronic systems, i.e. sensors, actuators, an ECU and the circuits in between them, because this feature helps to impose a structure on the study of vehicle electronics and fault diagnosis.

If any element of the system fails then the system itself will fail. For example, if a loose connection causes a break in the circuit between a sensor and the ECU the sensor data will no longer be transmitted to the ECU. The interdependence of the elements of a system needs to be understood so that a reliable procedure, such as the 'six step' approach, can be deployed.

So far we have identified two significant factors: one is a common basic structure for vehicle electronic systems and the other is an organised (six step) approach for diagnosis.

We will put the six steps to one side for the time being and concentrate on the elements of the system, for it is knowledge of this that will permit us to deploy the six step approach satisfactorily.

In the next chapter we will look at a range of examples of sensors, actuators, etc. that make up typical vehicle systems. We will follow that up with a study of the electrical and electronic principles that are known to be helpful to people who perform good diagnostic work on vehicle systems.

2
Common features of vehicle systems

Having looked at the basic structure of typical vehicle systems, this is a convenient point at which to consider some of the elements in a little more detail. Figure 2.1 reminds us of the basic elements of a system.

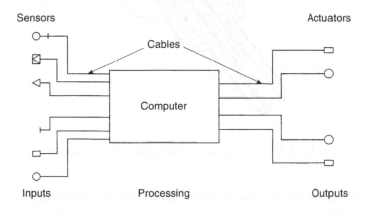

Fig. 2.1 The 4 elements of the system (sensors, actuators, circuits, and an ECU)

By taking a look at samples of each of the elements we shall obtain an indication of the type of electronics knowledge that is needed. The treatment assumes a knowledge of principles, such as the effects of an electric current, e.g. electro-magnetic effects, circuit diagrams and the symbols used. Electronic detail is not discussed here because this is dealt with in another chapter.

Basic components of a vehicle electronic system

SENSORS

A sensor is a device that 'senses' some physical quantity and produces an electrical response which can be made to represent that quantity. For example, the temperature of the engine coolant, the speed of a road wheel relative to the back plate of a brake, as in anti-lock brakes, and so on.

A COOLANT TEMPERATURE SENSOR

In keeping with the procedure adopted in Chapter 1 I will start with an engine coolant temperature sensor.

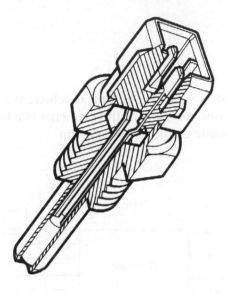

Fig. 2.2   An engine coolant temperature sensor

A commonly used device for sensing temperature is the thermistor. A thermistor utilises the concept of negative temperature coefficient. Most electrical conductors have a positive temperature coefficient. This means that the hotter the conductor gets the higher is its electrical resistance. This thermistor operates differently; its resistance gets lower as its temperature increases. There is a well-defined relationship between temperature and resistance. This means that current flow through the thermistor can be used to give an accurate representation of temperature.

Figure 2.3 shows the approximate relationship between temperature and resistance. The coolant temperature sensor provides the ECU with information about engine temperature and thus allows the ECU to make alterations to fuelling for cold starts and warm-up enrichment.

The information shown in Figure 2.3 may be given in tabular form as shown in Figure 2.4. This table shows the approximate resistance to be expected between the sensor terminals for a given temperature. From this it will be seen that it is possible to test such a sensor, for correctness of operation with the aid of a thermometer and a resistance meter (ohm-meter) provided the exact reference

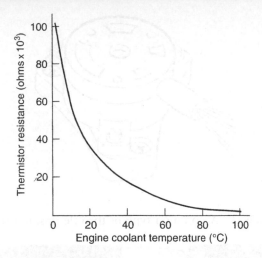

Fig. 2.3 Temperature versus resistance characteristics (thermistor)

Fig. 2.4 Table of temperature and corresponding resistance for a coolant sensor

Temp °C	Resistance kohm
20	60
30	32
60	10
80	4

values are known. Tools and equipment for use in electronic system diagnosis are covered in a later chapter, but at this point it should be noted that good quality, accurate meters of the correct resistance rating are the ones to use. Test lamps of the 12 V type are *not* suitable.

THROTTLE POSITION SENSOR

The throttle position sensor provides the ECU with data about throttle position and rate of movement. This enables the engine to respond instantly to the driver's throttle action and run smoothly under acceleration. The actual device is quite sophisticated but, like the coolant sensor, it operates according to well-established electrical principles.

Figures 2.6 and 2.7 show the basic principle of the throttle sensor. The wiper (moving part) of the potentiometer connects to one end of the throttle spindle. An electrical connection is made to a suitable point on the wiper and the voltage V_p measured, at this point, provides the signal that informs the computer about throttle position and movement.

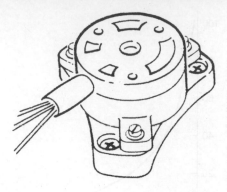

Fig. 2.5(a)   A common type of throttle position sensor

Fig. 2.5(b)   A throttle sensor on an engine

Here, as with the coolant sensor, it is evident that quite basic electrical test equipment, such as a voltmeter, will permit one to test for correctness of operation. Again such testing is subject to availability of precise data such as voltage for a given throttle position.

THE CRANKSHAFT ANGLE SENSOR (ENGINE SPEED AND PISTON POSITION)

The crank position sensor generates a signal to say where the piston is, e.g. top dead centre, 40° before TDC, etc., and speed of rotation of the crank, i.e. engine revs per minute (rpm). The crank sensor is sometimes located near the engine flywheel, so that signals are generated direct, but it may also be located at the camshaft or in the ignition distributor. Wherever it is situated, it serves much the same purpose.

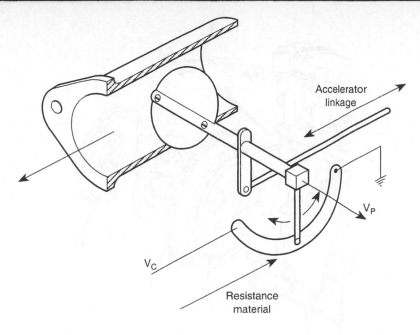

Fig. 2.6 Potentiometer type throttle position sensor

V_C = Constant voltage supply from computer

V_P = Voltage giving position of throttle

V_C = 5 V

V_P Varies with throttle angle

Fig. 2.7 The potential divider principle of the throttle position sensor

The type of sensor shown in Figure 2.8 relies for its operation on the difference in magnetic reluctance that is displayed by iron and air. In very general terms, reluctance is to magnetism what resistance is to electricity. Air has a high reluctance and iron a low reluctance. When the iron segments (poles) are aligned with the armature of the sensor a strong magnetic flux flows and the voltage induced at the sensor terminal is high; when the air gap is fully aligned with the sensor armature the magnetic flux is zero and the sensor voltage falls to zero. As the flywheel rotates the sensor produces a voltage of alternating form and, here again, if the values are known it is fairly easy to see that a suitable voltmeter

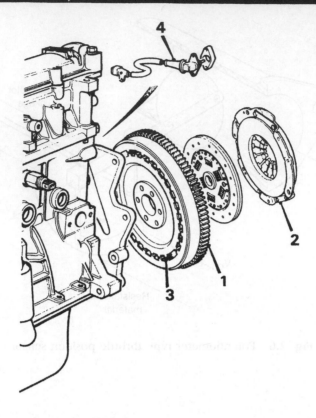

Fig. 2.8   Crank position and engine speed sensor. 1. Flywheel; 2. Clutch pressure
plate; 3. Reluctor disc; 4. Crank sensor and cable

applied at the sensor terminals, accompanied by activation of the sensor by
rotation of the flywheel, would provide an indication of sensor performance.

Figure 2.10 shows how the sensor voltage varies as one pole piece on the
reluctor rotates past the sensor armature.

The examples of sensors examined here are representative of many others that
are used on vehicles. It should be appreciated that they rely on basic physical
principles for their operation, and knowledge of these basic principles is assumed
for the purposes of this short survey.

In addition to making this point, I should mention that these sensors generate
the information (inputs) that enable the ECU to perform its work. In many cases
the 'raw' signal from a sensor requires electronic processing before it can be used
by the ECU, which is why we will look at some electronics in the next
chapter.

ACTUATORS

An actuator may be an electro-mechanical device. Petrol injectors are a common
form of actuator. Electrical pulses, of a given duration, energise the solenoid so

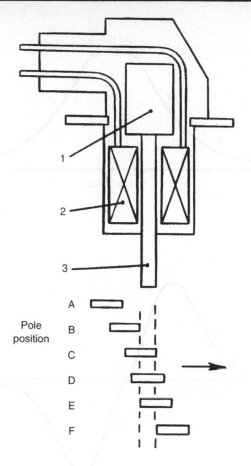

Fig. 2.9   Sensor details. 1. Permanent magnet; 2. Coil (voltage induced in here); 3. Armature assembly

that petrol is sprayed into the incoming air stream to provide the correct mixture to the cylinders of the engine at a frequency that is consistent with the engine speed. Figures 2.11 and 2.12 show typical petrol injectors as made by Lucas.

The electrical pulses from the ECU that are used to energise the injectors are passed through the solenoid winding. This action pulls the solenoid plunger away from the injector valve seat, against a spring. In this type of injector the valve lift is approximately 0.15 mm. The duration of injection, per pulse, varies in the region of 1.5 to 10 milliseconds.

A guide to the condition of an injector may be obtained by measuring the resistance of the solenoid winding. This is far from a complete check, but we will examine testing in more detail later.

A CRUISE CONTROL ACTUATOR

Cruise control automatically adjusts throttle position, so that a predetermined speed selected by the driver can be maintained in suitable driving conditions.

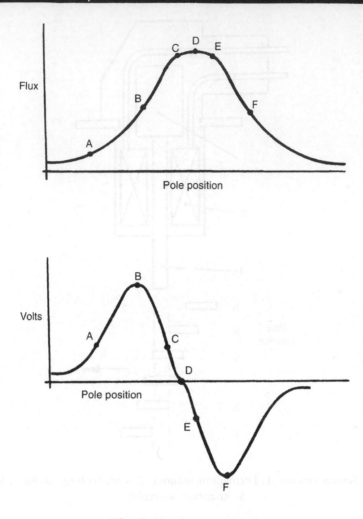

Fig. 2.10 Sensor voltage

Figure 2.13 shows the general arrangement on the car and Figure 2.14 shows a simplified picture of the actuator.

From Figure 2.14 it may be seen that there is a small solenoid which is under the control of the cruise control computer. There is also a toggle valve which is activated, by the solenoid, against the spring. In the position shown atmospheric air is entering and this helps to close the throttle more quickly. When the solenoid is energised the toggle valve is pulled towards the solenoid. This closes the atmospheric air valve and opens the vacuum valve and the diaphragm will cause the throttle to open. Here again, it is seen that the technology deployed, in the actuator, is based on well-established electro-mechanical principles.

The point I am making is that a thorough understanding of basic vehicle technology is an essential ingredient of the knowledge that vehicle repairers need if they are to perform good, accurate work on electronic systems.

Fig. 2.11   Petrol injector (Lucas)

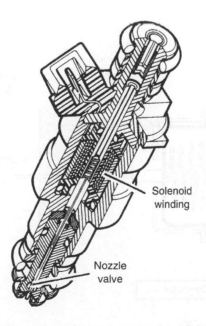

Fig. 2.12   Petrol injector constructor details (Lucas)

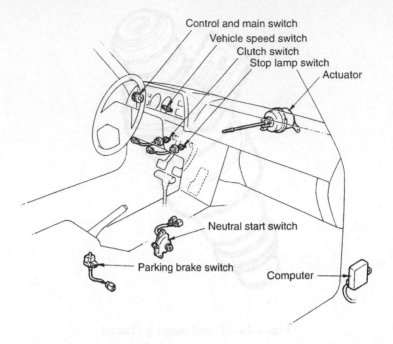

Fig. 2.13   Components of a cruise control system (Toyota)

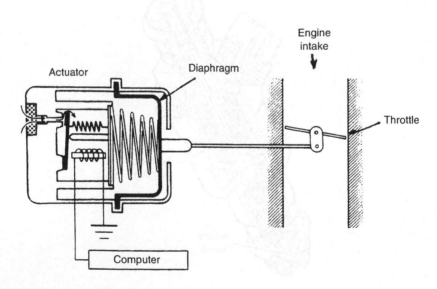

Fig. 2.14   A cruise control actuator (Toyota)

STEPPER MOTOR

The stepper motor is often used in electronically controlled systems. Figures 2.15(a) and (b) show the basic principle of a variable reluctance stepper motor. The rotor is constructed from soft iron and has six equally spaced lugs protruding from it. The soft iron rotor is mounted on a shaft between bearings. Surrounding the rotor, with an air gap between them, is the stator. This stator has eight

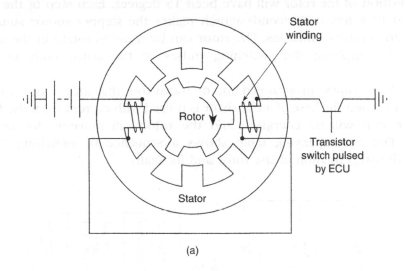

(a)

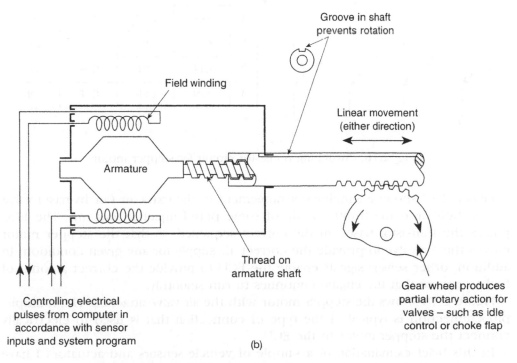

(b)

Fig. 2.15(a) and (b) Variable reluctance stepper motor

protrusions (lugs). Opposite pairs of stator lugs carry coils of wire through which current is passed to form a magnetic field.

The soft iron rotor is influenced by this magnetic field and it will rotate with its shaft until the rotor lugs line up with a pair of stator lugs, as shown. The motor shown has four pairs of stator lugs and three pairs of rotor lugs.

If one stator coil is switched off and the adjoining pair is switched on the rotor will rotate until the next set of rotor and stator lugs are aligned. In this case the angular motion of the rotor will have been 15 degrees. Each step of the rotor is performed in a few milliseconds which makes the stepper motor suitable for some engine control functions. The rotor can be made to rotate in the opposite direction by applying the switching pulses to the stator coils in reverse sequence.

Figure 2.16 shows an arrangement of transistors that are in the stator coil circuits and when the base of a transistor is pulsed with current from the ECU the respective coil will be energised and the rotor will move to its designated position. The accompanying table shows a sequence of switching that will produce the step rotation of the rotor and its shaft.

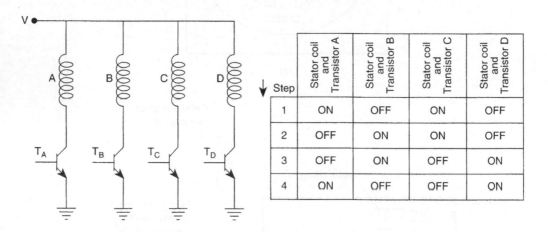

Step	Stator coil and Transistor A	Stator coil and Transistor B	Stator coil and Transistor C	Stator coil and Transistor D
1	ON	OFF	ON	OFF
2	OFF	ON	ON	OFF
3	OFF	ON	OFF	ON
4	ON	OFF	OFF	ON

Fig. 2.16 Switching transistor circuit for stepper motor

Figure 2.17 shows a simplified arrangement of the extra air (air by-pass) valve that is built into the throttle body of some petrol injection systems. The ECU pulses the transistor bases, in the correct sequence, so that the stepper motor moves the air valve to provide the correct air supply for any given condition. In addition, other sensor signals enable the ECU to provide the correct amount of fuel to ensure that the engine continues to run smoothly.

Figure 2.18 shows the stepper motor with the air valve attached. The multiple-pin connection is typical of the type of connection that is used to electrically connect the stepper motor to the ECU.

In this brief examination of a sample of vehicle sensors and actuators I have shown that they are based on fairly conventional technology and they are usually

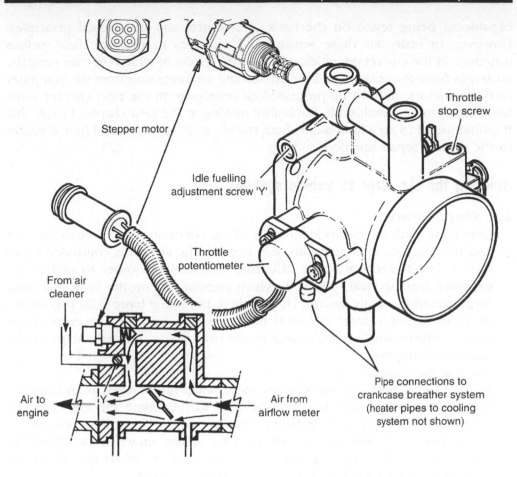

Stepper motor

Idle fuelling
adjustment screw 'Y'

Throttle
potentiometer

From air
cleaner

Throttle
stop screw

Air to
engine

Y

Air from
airflow meter

Pipe connections to
crankcase breather system
(heater pipes to cooling
system not shown)

Fig. 2.17 A stepper motor operated air valve

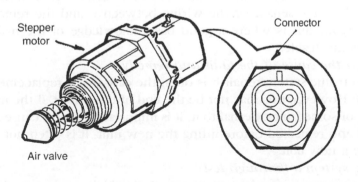

Stepper
motor

Connector

Air valve

Fig. 2.18 The stepper motor and extra air valve

capable of being tested on the basis of electrical and mechanical principles. However, in order for these sensors and actuators to perform their various functions in the correct operation of the entire system of which they are part, the messages from the sensors, and the controlling messages sent from the computer to the actuators, need to be processed electronically. In the next chapter some basic electronics is dealt with, but before moving to the next chapter I think that it would useful to say a little more about the 'six step' approach and how it relates to the vehicle repair situation.

Relating the six steps to vehicle systems

1. *Collect evidence*
   Collecting evidence means looking for all the symptoms that relate to the fault and not jumping to conclusions, e.g. because the system is controlled by an ECU it is not necessarily the ECU that is at fault. In order to collect the evidence it is necessary to know which components on the vehicle actually form part of the faulty system. This is where the sound basic skills part comes in. If an engine control system is malfunctioning because one cylinder has poor compression it is important to discover this at an early stage of the diagnostic process.

2. *Analyse the evidence*
   In the case of poor compression on one cylinder given above as an example, the analysis would take the form of tests to determine the cause of low compression, e.g. burnt valve, blown head gasket, etc. The analysis of evidence that is performed will vary according to the system under investigation. But these steps are obviously important otherwise one may embark on a drawn-out electronics test procedure which will prove unproductive.

3. *Locate the fault*
   The procedure for doing this on an electronic system varies according to the type of test equipment available. It may be the case that the system has some self-diagnostics which will lead you to the area of the system which is defective. Let us assume that this is the case and the self-diagnostics reports that an engine coolant temperature sensor is defective. How do you know whether it is the sensor or the wiring between it and the remainder of the system? Again this is where a good basic knowledge of the make-up of the system is invaluable.

4. & 5. *Find the cause of the fault and remedy it*
   With electronic system repair it is often the case that a replacement unit must be fitted. However, this may not be the end of the matter. If the unit has failed because of some fault external to it, it is important that this cause of the failure is found and remedied before fitting the new unit. It is often not just a matter of fitting a new unit.

6. *Give the system a thorough test*
   Testing after repair is an important aspect of vehicle work, especially so where electronically controlled systems are concerned. In the case of intermittent

faults such testing may need to be extended because the fault may only occur when the engine is hot and the vehicle is being used in a particular way.

In this chapter, so far, we have considered some sensors and actuators. We have not covered the interconnecting circuits because these are dealt with later, in Chapter 6. The remaining element of our system is the ECU and this is an appropriate point at which to give some attention to this device.

The electronic control unit (ECU)

In this section the intention is to provide an insight into the way in which an electronic control unit (ECU) works. Seldom, if ever, would a vehicle technician in a garage have to attempt to repair a control unit. The normal process of diagnosis and repair requires the 'black box' treatment. That is to say that the defective unit must be accurately located, the cause of its failure rectified and then a *correct* replacement unit fitted. Most test equipment is made on the assumption that this is the case. The word 'correct' is emphasised because one ECU for a particular vehicle may look exactly like another, but the two units may have quite

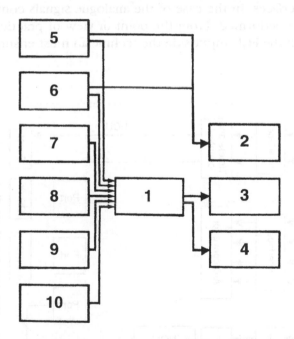

1. Electronic Control Unit
2. Fuel pump
3. Group 1 injectors
4. Group 2 injectors
5. Engine start relay
6. Engine air flow
7. Engine speed
8. Coolant temperature
9. Air temperature
10. Throttle potentiometer

Fig. 2.19 The ECU in the system

different programs stored in them. If an incorrect unit is fitted as a replacement there could be disastrous consequences, and so it is imperative that only the exact correct replacement unit is used. It is, therefore, believed that an insight into the operating principles of the 'black box' (ECU) can be helpful in developing an understanding of the method of operation of the entire system. Equally important, perhaps, is that intellectual curiosity may be satisfied.

The electronic control unit in a vehicle system may be called an ECU, an ECM (engine control module), a controller, a microprocessor or some other name. To a large extent, it depends on the system in which the device is used as a controller, and the make and age of the vehicle to which it is fitted. However, for the current purpose, I shall use the term electronic control unit (ECU), and work on the principle that it contains a microprocessor (or equivalent) and that the microprocessor operates on computing (digital) principles.

Figure 2.20 shows a simplified diagram of an ECU. The internal circuits (buses) that link the units together are not shown and the device is simplified to show the elements that are most important from the service technician's point of view.

The signal inputs that are obtained from the sensors on the vehicle are shown as pulse, analogue and digital. The 'raw' signals from the sensors require processing, i.e. amplifying, cleaning up (filtering), etc., and these processes are performed at the interfaces. In the case of the analogue signals conversion to digital form must also be performed. From the point of view of practical fault diagnosis this means that on the ECU input side the technician must ensure that the sensor values are correct.

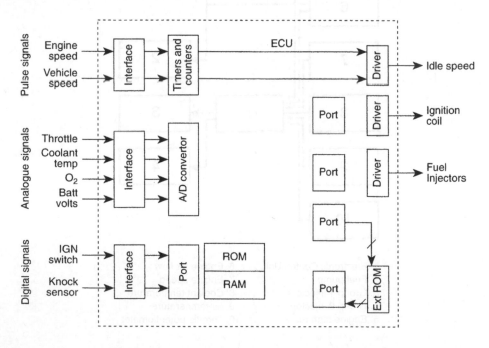

Fig. 2.20 A simplified microcontroller type ECU

On the ECU output side 'drivers' are used to convert the computer logic levels of the ECU into the higher power signals that are required to drive actuators, such as fuel injectors, extra air valve stepper motors, and, in the case of ABS, motors and valves. A range of interfaces, drivers, memories and other electronic devices is described in Chapter 8.

Modern ECUs usually include an external diagnostic (serial) port through which diagnostic equipment can direct the processor to perform functional checks of the input and output circuits. To aid this process a section of durable memory may be set aside to store fault codes that have been generated while the vehicle has been operating. The electronic memory used for this purpose may be an EEPROM which stores fault codes until they are removed by a specially generated voltage pulse, or it may be a semi-permanent memory KAM (keep alive memory), which is energised direct from the battery, not via the ignition switch. The information stored in such a memory can be read out through the self-diagnostics on the vehicle or by an instrument connected to the diagnostic port of the ECU. These topics are covered in more detail in Chapters 7 and 8.

Because it is impracticable for the average garage to test an ECU it is vitally important that all sensors and the ECU inputs that they generate, and all actuators that operate on ECU outputs, are properly verified before an ECU is changed. And, as stated earlier, it is also important to ensure that the defect is not caused by something simple such as a fouled spark plug, or a disconnected cable. This is why it is so important when working on electronically controlled systems to deploy an orderly strategy, such as the 'six steps'.

3
Introduction to electronics

The previous chapters have concentrated on describing the structure of vehicle electronic systems. This is now a convenient point at which to introduce some electrical revision and to consider some basic electronics.

Electrical principles

It is generally accepted that there are three effects of an electric current, i.e. heating, chemical and magnetic. For these three effects to occur an electric current must flow and this can only happen when there is a complete circuit.

HEATING EFFECT OF AN ELECTRIC CURRENT

When current flows in a headlamp bulb filament the heat causes the filament to become incandescent and to emit light.

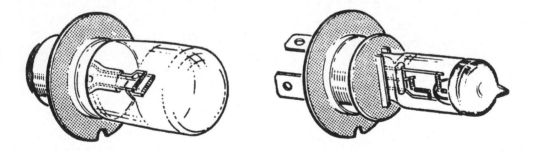

Fig. 3.1 Heating effect

CHEMICAL EFFECT OF ELECTRIC CURRENT

The vehicle battery is a device which utilises the chemical effect. It converts electrical energy into chemical energy through the action of electric current on

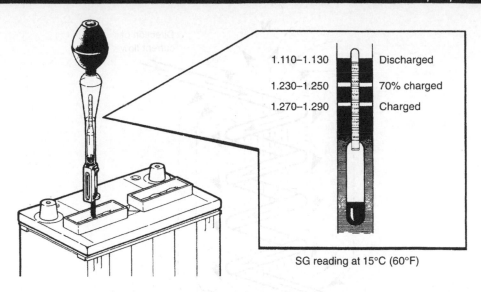

1.110–1.130	Discharged
1.230–1.250	70% charged
1.270–1.290	Charged

SG reading at 15°C (60°F)

Fig. 3.2   Vehicle battery – chemical effect

the battery plates and the electrolyte (dilute sulphuric acid). When the battery is required to provide energy to vehicle systems the chemical energy is converted back to electricity.

THE MAGNETIC EFFECT OF AN ELECTRIC CURRENT

Figure 3.3 shows how a circular magnetic field is set up around a wire (conductor) which is carrying electric current.

When a conductor (wire) is made into a coil the magnetic field created is of the form shown in Figure 3.4. A coil such as this is the basis of a solenoid.

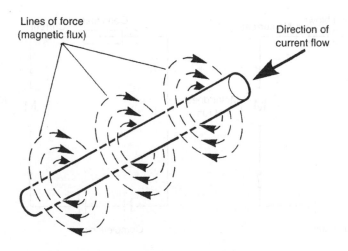

Fig. 3.3   Magnetic field around a straight conductor

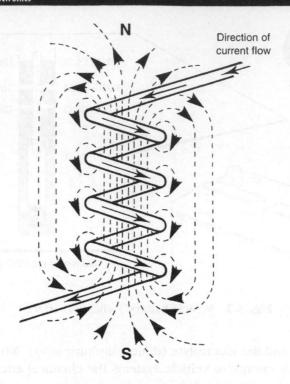

Fig. 3.4 Magnetic field of a coil

A complete circuit is needed for the flow of electric current. Figure 3.5(a) shows two diagrams of a motor in a circuit. Should the fuse be blown the circuit is incomplete; current will not flow and the motor will not run. In the right-hand diagram the fuse has been replaced. The circuit is now complete and the motor

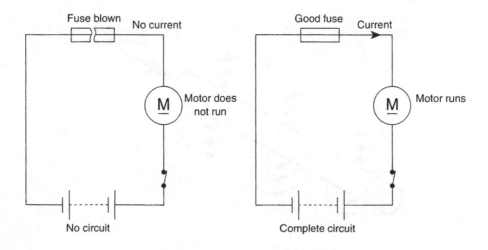

Fig. 3.5(a) An electric circuit

Name of device	Symbol	Name of device	Symbol
Electric cell		Lamp (bulb)	
Battery		Diode	
Resistor		Transistor (NPN)	
Variable resistor		Light emitting diode (LED)	
Potentiometer		Switch	
Capacitor		Conductors (wires) crossing	
Inductor (coil)		Conductors (wires) joining	
Transformer		Zener diode	
Fuse		Light dependent resistor (LDR)	

Fig. 3.5(b) A selection of circuit symbols

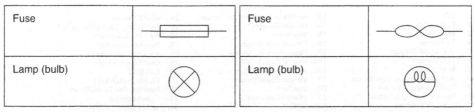

Fuse		Fuse	
Lamp (bulb)		Lamp (bulb)	

Approved symbol   **Symbol sometimes used**

Fig. 3.5(c) Non-standard circuit symbols

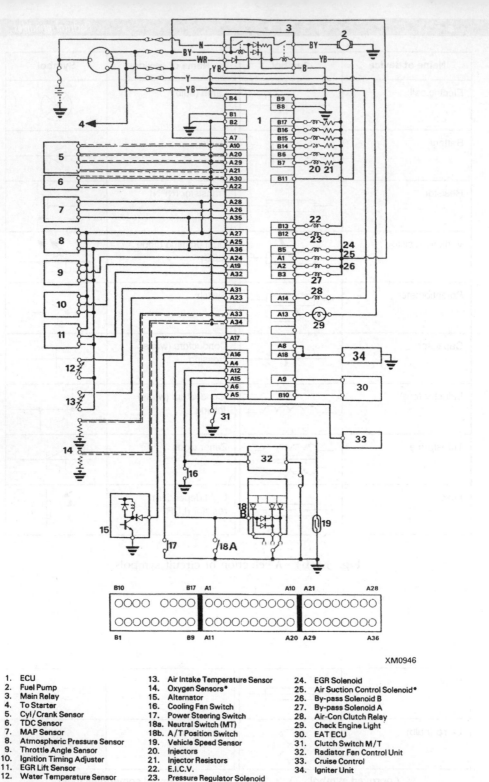

XM0946

1. ECU	13. Air Intake Temperature Sensor	24. EGR Solenoid
2. Fuel Pump	14. Oxygen Sensors*	25. Air Suction Control Solenoid*
3. Main Relay	15. Alternator	26. By-pass Solenoid B
4. To Starter	16. Cooling Fan Switch	27. By-pass Solenoid A
5. Cyl/Crank Sensor	17. Power Steering Switch	28. Air-Con Clutch Relay
6. TDC Sensor	18a. Neutral Switch (MT)	29. Check Engine Light
7. MAP Sensor	18b. A/T Position Switch	30. EAT ECU
8. Atmospheric Pressure Sensor	19. Vehicle Speed Sensor	31. Clutch Switch M/T
9. Throttle Angle Sensor	20. Injectors	32. Radiator Fan Control Unit
10. Ignition Timing Adjuster	21. Injector Resistors	33. Cruise Control
11. EGR Lift Sensor	22. E.I.C.V.	34. Igniter Unit
12. Water Temperature Sensor	23. Pressure Regulator Solenoid	

* Emission Vehicles Only

Fig. 3.5(d)   System diagram with list describing circuit elements

will run. Basic electrical principles such as this are fundamental to good work on electronic systems because much of the testing of such systems requires checking of circuits to ensure that they are complete.

It is essential to be able to understand and follow circuit diagrams, and this requires a knowledge of circuit symbols. There is a set of standard circuit symbols, some of which are shown in Figure 3.5(b). However, non-standard symbols are sometimes used and this can cause confusion. Figure 3.5(c) shows an alternative symbol for a lamp (bulb) and one for a fuse. Fortunately, when other symbols are used in a circuit diagram they are usually accompanied by a descriptive list. Figure 3.5(d) gives an example: the injector resistors are shown as a saw-tooth line, number 21 on the list, and on the diagram. This solves the problem of deciphering the diagram.

The remainder of treatment in this book assumes that the reader has a knowledge of electrical principles obtained by study of City & Guilds Course 383 'Road Vehicle Science Background', or an equivalent level of knowledge. Basic electrical knowledge is an essential ingredient of the skill needed for accurate work on vehicle electronic systems and most of it can be learned by building practical electrical circuits and observing inputs and outputs; doing it this way overcomes the maths problem that sometimes deters practical people.

Electronics

Electronics is a branch of electrical engineering and it is concerned with 'the science and technology of the conduction of electricity in a vacuum, a gas or a semi-conductor'. It is the semi-conductor that concerns us most, at this stage, so it is necessary to explain what the term means.

SEMI-CONDUCTORS

Metals such as copper, aluminium and gold are good conductors of electricity; other materials like rubber and PVC are bad conductors and they are known as insulators. Materials such as silicon and germanium have conductivity which lies between that of good conductors and insulators. These materials are semi-conductors. Semi-conductors allow an electric current to flow only under certain circumstances.

DOPING

Pure silicon, when treated (doped) with substances such as aluminium or gallium, is widely used in the manufacture of electronic components. The doping agents, which are added in tiny concentrations, probably less than one part per million, cause the silicon to become more conductive. The doping agents are divided into two groups: 1. acceptors; and 2. donors.

Typical acceptors are boron, gallium, indium and aluminium. When an acceptor is added to the silicon it forms p-type silicon. The acceptor forms holes which behave like a mobile positive charge and so this silicon is called p-type.

Typical donors are arsenic, phosphorous and antimony. When a donor is added to the silicon it forms n-type silicon. It is called n-type because electric currents in it are carried by (negatively charged) electrons.

THE P–N JUNCTION

A p–n junction is a junction between p-type silicon and n-type silicon, within a single crystal. If wires are attached to the material on either side of the p–n junction it is found that current will pass through the device in one direction but hardly at all in the opposite direction. Such a device is a junction rectifier which is commonly known as a junction diode, or just 'diode'. We thus have a device that acts like a one-way valve, i.e. it will pass current in one direction but not the other.

Figure 3.6 shows a representation of a p–n junction, in silicon, connected to a source of electricity. When it is connected in the way shown, with the positive battery terminal connected to the p-type material, the junction is said to be forward biased. When the forward voltage reaches approximately 0.7 V a continuous current will flow in the forward direction. Diodes are made for many applications such as rectification of a.c. to d.c., switching circuits and so on. Figure 3.7 shows a diode in a metallic heat sink. The second terminal of the diode

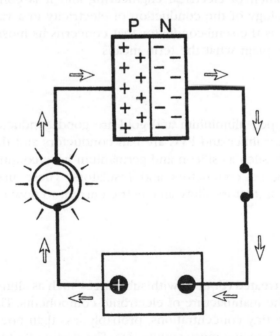

Fig. 3.6 A p–n junction

Diode

Metallic
heat sink

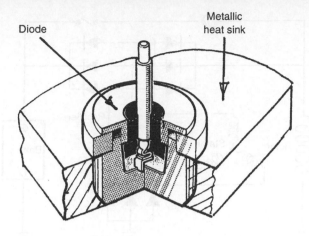

Fig. 3.7   A diode

Fig. 3.8   Circuit symbol for a diode

is formed from the metal case into which it is built, and the heat sink conducts heat away from the diode.

Figure 3.8 shows the symbol for a diode. The arrow indicates the direction in which current will flow. Diodes are used in circuits to permit current flow in one direction only.

A common use of a diode is in the conversion (rectifying) of alternating current (a.c.) to direct current (d.c.), as in a vehicle alternator. The rectifier circuit shown in Figure 3.9 is quite complicated and it is shown just to illustrate an application of diodes in vehicle practice.

Diodes come in many forms and are one of the basic building blocks of electronic circuits.

ANOTHER ELECTRONIC USE OF THE P-N JUNCTION

The transistor is another basic building block of electronic circuits. It may be thought of as two p-n junction diodes, connected back to back, as shown in

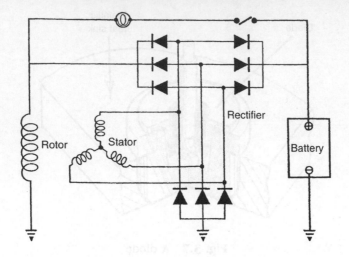

Fig. 3.9 A rectifier circuit

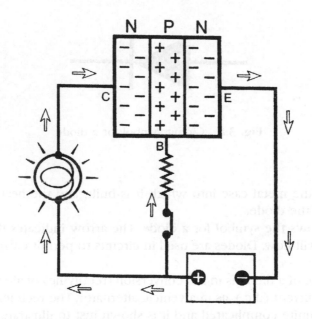

Fig. 3.10 Two p-n junctions back to back

Figure 3.10. Transistors are used as switches or current/voltage amplifiers, and in many other applications.

Transistors have three connections, named as follows: Base – (B); Collector – (C); Emitter – (E).

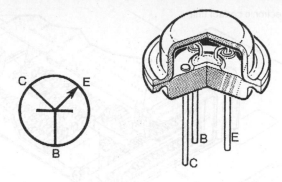

Fig. 3.11 A transistor

Transistors are non-conductive until a very small current flows through the base-emitter circuit. This very small current turns on the collector-emitter circuit and permits a larger current to flow through the collector/emitter circuit. When the base-emitter current is switched off the collector-emitter ceases immediately to conduct and the transistor is effectively switched off. This happens in a fraction of a second (a few nano-seconds usually) and makes the transistor suitable for high-speed switching operations, as required in many vehicle applications.

INTEGRATED CIRCUITS (I/Cs)

Many diodes, resistors and transistors can be made on a single piece of silicon (a chip). When the diodes, resistors and transistors are connected together, in specific ways, to make circuits on one piece of silicon, the resulting device is

Fig. 3.12 A typical I/C

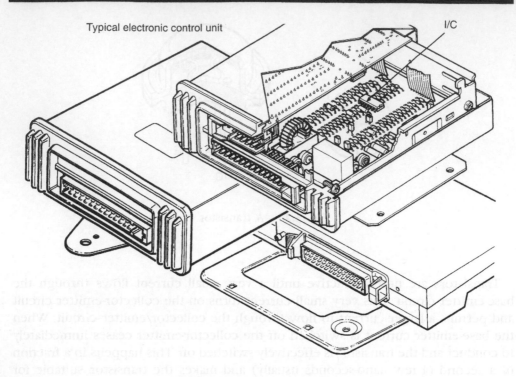

Typical electronic control unit   I/C

Fig. 3.13   An I/C in an ECU

known as an integrated circuit (I/C). When a small number of diodes, transistors and resistors are connected together (about 12) the result is called small-scale integration (SSI); when the number is greater than 100 it is called large-scale integration (LSI). From the outside an I/C looks like Figure 3.12.

The 'legs' along the sides are metal tags which are used to connect the I/C to other components. The number of tags is related to the complexity of the I/C. When I/Cs are packaged, as shown, they are known as DIL type, which means that the 'legs' are dual in line. Figure 3.13 shows an I/C built into an ECU.

How do electronic devices work?

If we now look at a small circuit from a vehicle system we should get an idea of the type of theory that we need to come to terms with, in order to understand how the system works.

Alternator charging systems have been in use for many years, and their rectifier circuits and voltage regulator circuits incorporate transistors and diodes, so this is probably a good example to use for our current purpose – which is to learn a little more about transistors and diodes.

THE ALTERNATING CURRENT GENERATOR (ALTERNATOR)

Fig. 3.14 A typical alternator

THE RECTIFIER

To keep the vehicle battery charged a direct current (d.c.) supply is required. The basic product of an alternator is alternating current (a.c.). One of the first jobs of electronics is to convert the a.c. into d.c., and this is performed by the rectifier which is normally incorporated into the alternator assembly. The use of diodes in a rectifier was referred to in Figure 3.9.

A half-wave rectifier
The bridge rectifier is quite a complicated circuit so we will start with a simpler one called a half-wave rectifier.

Reference to Figure 3.15 shows that the diode permits the half wave that is positive to flow through the circuit, but prevents the negative half wave, between 180 and 360 degrees, from flowing. The result is that alternate half waves of a.c. at the rectifier terminals will cause a current to flow, in one direction, through the resistor R. Such a rectifier can be made to produce useful d.c. output by the addition of a smoothing capacitor to the circuit. However, we are concentrating on the diode action so we will not consider the capacitor at this stage.

As stated, the diode conducts when the p–n junction has a positive voltage of about 0.7 V applied to the p terminal (forward biased). When the voltage drops below this value, and when it becames negative, the diode ceases to conduct. The diode thus makes an effective one-way valve for electric current, and when several diodes are arranged in a circuit it is possible to obtain full-wave rectification, which is what is used in vehicle charging systems.

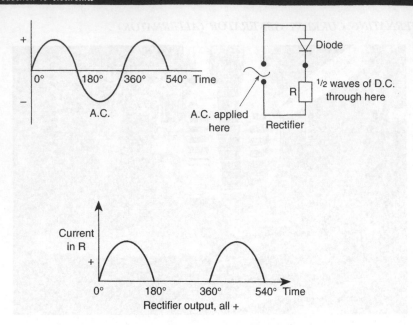

Fig. 3.15 A half-wave rectifier

A full-wave rectifier

To obtain a steady d.c. supply for battery charging and operation of vehicle systems it is necessary to use a full-wave rectifier. The particular one shown in Figure 3.16 uses a circuit layout known as a bridge circuit. This is similar to the ones used on vehicles and it serves our current purpose, which is to develop some understanding of diodes and their uses.

Figure 3.16(a) shows a full-wave rectifier circuit. There are four diodes, A, B, C and D, which are connected in bridge form. There is an a.c. input and a voltmeter is connected across the output resistor.

Reference to (b) shows how the current flows for one half of the a.c. cycle, and (c) shows the current flow in the other half of the a.c. cycle. In both cases the current flows, through the load resistor R, in the same direction. The output is thus direct current and this is achieved by the 'one way' action of the diodes.

In operation a power loss occurs across the diodes and in order to dissipate the heat generated the diodes are mounted on a heat sink which conducts the heat away into the atmosphere. In Figure 3.17 the heat sink is formed by the metal plate which holds the rectifier diodes.

DIODE BEHAVIOUR

The forward biased p-n junction has the positive of the supply connected to the p side and when the voltage reaches approximately 0.7 V the p-n junction diode conducts electricity.

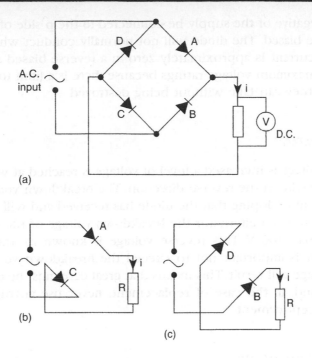

(a)

(b)

(c)

Fig. 3.16 A full-wave rectifier

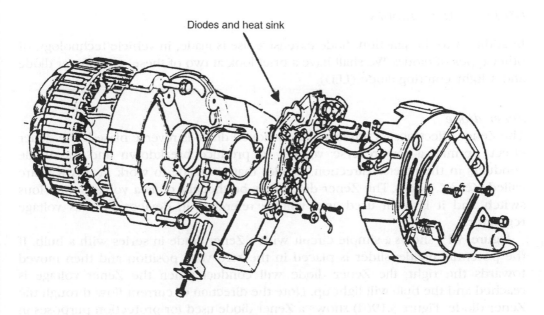

Diodes and heat sink

Fig. 3.17 Metal plate heat sink

Should the negative of the supply be connected to the p side of the junction it becomes reverse biased. The diode will not normally conduct when it is reverse biased (strictly current is approximately zero in a reverse biased silicon diode).

Diodes have maximum voltage ratings because there is a limit to the amount of reverse voltage they can take without being destroyed.

BREAKDOWN VOLTAGE

If the reverse voltage is increased a level of voltage is reached at which the diode will conduct, heavily, in the reverse direction. The breakdown voltage is dependent on the amount of doping that the diode has received and will vary according to the application; in an alternator the breakdown voltage of the rectifier diodes may be more than 100 V. This reverse voltage is known as 'the peak inverse voltage' (PIV). It is important not to exceed the breakdown voltage otherwise permanent damage will result. This means that great care must be exercised when testing diodes and, in the case of replacement, never use anything except the exactly correct replacement.

TESTING A JUNCTION DIODE

The fact that a diode conducts freely in the forward direction and virtually not at all in the reverse direction means that an ohm-meter can be used to check for correct operation of the diode.

OTHER TYPES OF DIODES

In addition to the junction diode extensive use is made, in vehicle technology, of other types of diodes. We shall have a brief look at two of these – the Zener diode and a light emitting diode (LED).

Zener diode

The Zener effect occurs when a heavily doped diode is reverse biased. The Zener effect permits a low reverse voltage to produce breakdown and the diode conducts in the reverse direction. Diodes that are made to work in this way are called 'Zener' diodes. The Zener diode may be thought of as a voltage conscious switch and it is often used as a voltage reference in devices such as voltage regulators.

Figure 3.19 shows a simple circuit with a Zener diode in series with a bulb. If the potential divider slider is placed in the zero volts position and then moved towards the right, the Zener diode will conduct when the Zener voltage is reached and the bulb will light up. Note the direction of current flow through the Zener diode. Figure 3.19(d) shows a Zener diode used for protection purposes in an alternator circuit.

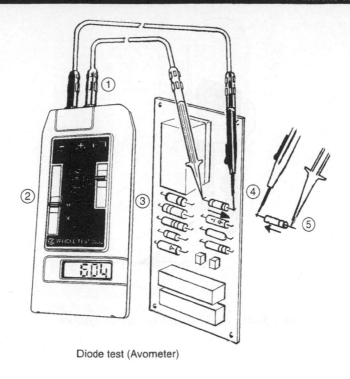

Diode test (Avometer)

Fig. 3.18 Diode test using Avo 2002 (Thorn EMI)

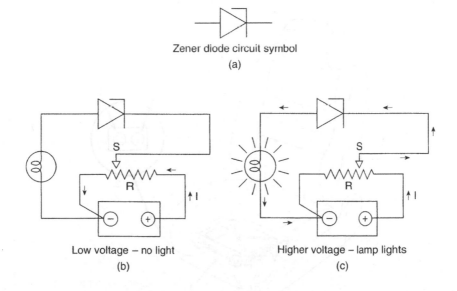

Zener diode circuit symbol
(a)

Low voltage – no light
(b)

Higher voltage – lamp lights
(c)

Fig. 3.19 Action of the Zener diode

A light emitting diode (LED)

Silicon is an opaque material which blocks the passage of light and the energy which derives from the action of the diode is given off as heat. LEDs are different

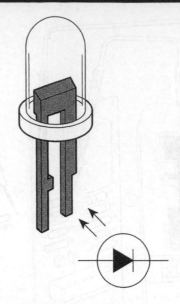

Fig. 3.20 An LED and its circuit symbol

1. Electronic control unit   3. Yellow LED
2. LED display   4. Red LED

Fig. 3.21 LEDs for fault code display

because they use materials such as gallium, arsenic and phosphorous, and the energy deriving from their operation is given out as light. The material from which the LED is made determines the colour of the light given out. It may be considered that the LED is a semi-conductor device which gives out light when current flows through it. LEDs are used, for example, as indicator lights and for fault indicator codes on some electronic systems.

The transistor

The transistor is another of our basic building blocks and as in the case of the diode we can take another section of a vehicle system and use it to develop some understanding of the operation and uses of transistors. However, we first need to examine the way in which transistors operate.

A transistor has three connections: the base, the emitter and the collector. When the transistor is connected into a circuit a very small current in the base to emitter circuit permits a very much larger current to flow in the collector-emitter circuit.

There are three parts to Figure 3.22. Part 1 shows the transistor as made from two p–n junctions (back-to-back). Part 2 shows the transistor symbol. Part 3 shows the construction of a fairly heavy duty transistor.

I wish to concentrate on Part 1 in order to describe the operation. There are two basic circuits here: one from the battery to the bulb and then to the collector terminal C of the transistor. When the transistor is conducting this circuit will be completed to the emitter E, and then back to the battery negative.

The other circuit shows the battery positive connected to the base terminal B of the transistor via a switch and a resistor. When the base voltage reaches a certain value current will flow from the base B to the emitter E and back to the battery negative.

A very small current in the base-emitter circuit permits a much larger current to flow in the collector-emitter circuit and the bulb will light up. If the switch in the base-emitter circuit is opened the transistor will cease to conduct from C to E and the bulb will go out.

If the base-emitter current is provided by electrical pulses of short duration, instead of the battery supply shown here, the light will go on and off at the same frequency as the electrical pulses.

USING THE TRANSISTOR AS A SWITCH

We have now added the transistor to our kit of basic building blocks and we should be able to understand how the transistor can be used to 'make and break' the primary circuit of an electronic ignition system.

Figure 3.23(a) shows a Hall effect ignition system. It is called a Hall type because the base current for the switching transistor TR2 is produced by a sensor (3.23(b)) which utilises the Hall effect. At this stage, I wish to concentrate on TR2, which is in the amplifier box in 3.23(a).

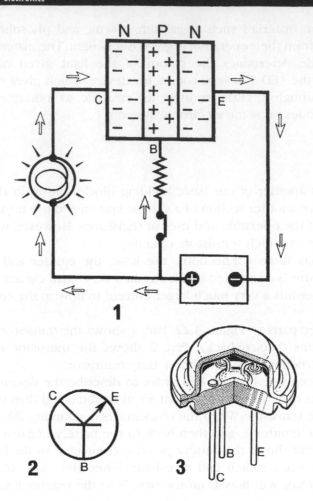

Fig. 3.22 Action of the transistor as a switch

The raw output from the Hall sensor is not suitable for direct use in operating TR2 and it is necessary to introduce a switching and smoothing circuit. The output from the switching and smoothing circuit provides the base current for TR2.

Note that there are two circuits in TR2: the collector-emitter circuit, which is in the primary circuit of the HT coil, and the base-emitter circuit which receives electrical pulses from the switching and smoothing circuit. These electrical pulses occur at the same frequency as is required for sparks at the sparking plugs. So every time a spark is required TR2 is switched off by the removal of an electrical pulse from the base of TR2.

Up to this point we have examined the diode and its use in an alternator circuit, and we have discussed some of the properties of transistors. Before proceeding to apply these principles to a study of a sample of electronically controlled systems I think that it is reasonable to consider some practical diagnostic tests that can be

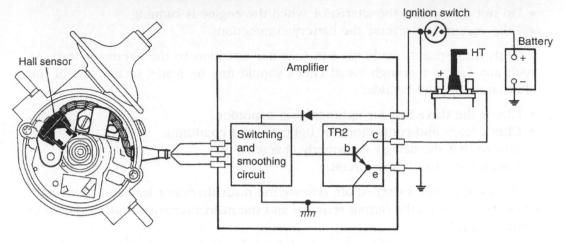

(a)   Lucas 'Hall effect' distributor coupled to a block diagram of the amplifier

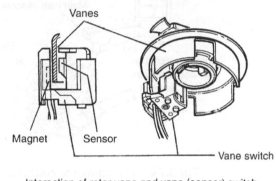

(b)   Interaction of rotor vane and vane (sensor) switch

Fig. 3.23(a) and (b)   A Hall type electronic ignition system (Lucas)

used on charging systems. This should help to consolidate the work covered so far and provide further confidence when readers move on to the following sections.

Alternator testing

Before starting on the alternator test procedure there are a few general points about care of alternators that need to be made.

- Do not run the engine with the alternator leads disconnected. (Note that in the tests that follow the engine is switched off and the ammeter is connected before restarting the engine.)
- Always disconnect the alternator and battery when using an electric welder on the vehicle. Failure to do this may cause stray currents to harm the alternator electronics.

- Do not disconnect the alternator when the engine is running.
- Take care not to reverse the battery connections.

With these points made we can turn our attention to the alternator tests. As with any test, a thorough visual check should first be made. In the case of the alternator this will include:

- Check the drive belt for tightness and condition.
- Check leads and connectors for tightness and condition.
- Ensure that the battery is properly charged.
- Check any fuses in the circuit.

With the engine switched off, remove the main alternator lead and connect an ammeter between the output terminal and the main alternator lead, as shown in Figure 3.24.

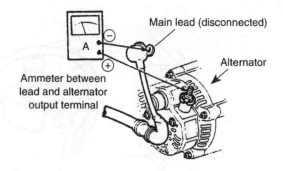

Fig 3.24 The ammeter connections

Once the ammeter connections are correctly made and the meter is placed in a secure position, where it is not likely to fall or foul any moving parts, the engine may be started.

To place a load on the alternator switch on the headlights, the rear screen heater and the air blower motor. Operate the accelerator pedal and observe the ammeter reading as the engine speed is increased. The output (for this particular alternator) should fall within the limits shown on the output graph, as shown in Figure 3.25.

If the alternator output is incorrect it will be necessary to remove the alternator so that it can be opened up for further testing and examination. Naturally, it will be necessary to clear the test meter away and to disconnect the battery before commencing to remove the alternator.

When the alternator is removed and cleaned ready for dismantling it should be placed on a clean work surface where there is plenty of space to lay out the dismantled parts.

When the alternator has been dismantled, the brushes and brush springs should be examined. The brushes should not be less than the recommended length and

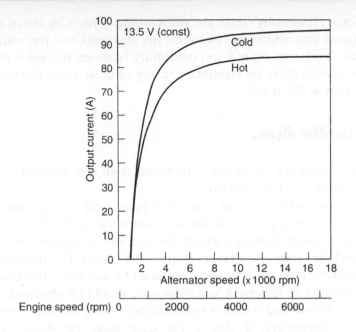

Fig 3.25 Alternator current verses engine RPM

they should move freely in the brush holders. The springs should be able to exert a force of 2 to 3 newtons (7 to 12 ounces) when compressed (depending on the type of alternator).

The rotor slip rings should be examined and if necessary cleaned with fine, clean glass paper. The multi-meter should now be switched to the ohms range and the resistance tests shown in Figure 3.26 applied to the rotor.

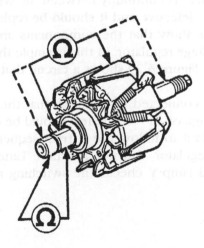

Fig 3.26 Testing the rotor

The resistance (continuity) tests are shown here. One is to check that there is infinite resistance (no continuity) between the slip rings and the rotor shaft, and the other is to check that there is no continuity between the rotor elements and the slip rings. Should there be continuity in any of these tests, the recommended remedy is to renew the rotor.

Testing the rectifier diodes

Figure 3.27(a) shows the diode pack removed from the alternator, and 3.27(b) shows the rectifier diode connections.

Because a diode (subject to the correct voltage) will only conduct current in one direction, a continuity test with an ohm-meter will verify the condition of the diodes. The multi-meter batteries should be checked to ensure that they are not discharged, and the meter must be set to the ohms range. By connecting one lead of the multi-meter to the main output terminal (1), and the other lead to each of the diode terminals (2), in turn, a low resistance should be obtained. If the meter leads are then reversed and the tests repeated each diode should show a very high resistance (no continuity). If this is the case then the diodes are working correctly.

If one multi-meter lead is now connected to earth terminal (3) the other four diodes may be checked by repeating the above procedure.

Testing the stator circuits

Figure 3.28 shows the ohm-meter connected to check the condition of the stator windings. There should be continuity between each pair of leads but not between the leads and the alternator casing. If the resistance of the coils falls outside the prescribed limits, or if there is continuity between the winding terminals and the casing, then the stator is defective and it should be replaced.

Should the above tests show that the components are in working order it is possible to check the voltage regulator. In this example the voltage regulator is an integrated circuit (chip). Figure 3.29 shows a circuit that can be used to test the regulator.

The two batteries are connected in series so that the variable potentiometer can provide the test voltage of 15+ V. The test should be carried out as quickly as possible and it is important to follow the correct sequence of switch operation otherwise the voltage regulator will not function. Lamp X checks the charge warning light circuit and lamp Y checks the switching action to the rotor field winding.

THE TEST PROCEDURE

With both switches in the 'OFF' position, slowly rotate the potentiometer knob to increase the voltage supply to the regulator to 12 V (observe the voltmeter reading

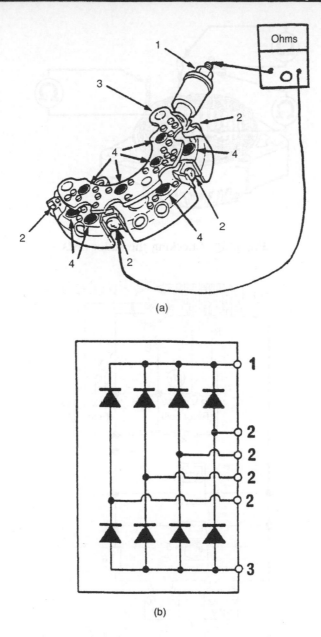

(a)

(b)

Fig 3.27 (a) Alternator details. 1. Main output terminal; 2. Diode terminals to stator; 3. Earth terminal; 4. Diodes. **(b)** Continuity test on alternator diodes

to ensure the correct direction). Turn switch 2 'ON' leaving switch 3 'OFF'. Lamps X and Y should light.

Turn switch 3 'ON' and leave switch 2 'ON'. Lamp X should switch off while lamp Y stays on.

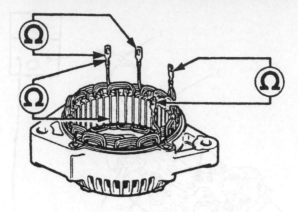

Fig 3.28 Checking the stator coils

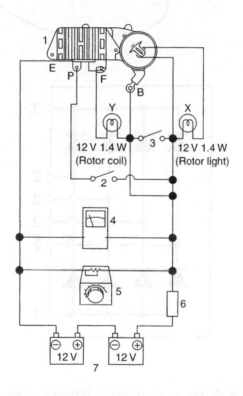

Fig 3.29 Testing an I/C type voltage regulator. 1. I/C regulator potentiometer; 2. Switch 2; 3. Switch 3; 4. Voltmeter; 5. Variable resistor; 6. 20 ohm resistor; 7. 12 V batteries

With both switches in the 'ON' position, slowly turn the potentiometer knob to increase the voltage supply to the regulator. Observe the voltmeter carefully as 14 volts is approached. Lamp Y should switch off at 13.9 V to 15 V. If the lamps do not light in the sequence described, the voltage regulator is faulty.

When all checks are completed, new parts fitted and the alternator rebuilt and refitted to the vehicle, a thorough test should be carried out to ensure that everything is working satisfactorily. Of course, if the alternator is seriously worn a decision will need to be made as to whether a replacement alternator might be the best option. That will be a matter of judgement.

The alternator is an example of the use of electronics on vehicles that has been in use for many years. As more complicated systems, such as engine control, have entered into use the electronic controller (ECU) has come to play a significant part in the operation of the system. This has led to the development of specialised tools for fault diagnosis purposes and the approach to repair and maintenance is somewhat different from the approach to alternator repair. Many sensors and actuators are not repairable in the garage and it is normally necessary to fit a replacement unit in order to make a repair. To assess the extent of this difference, the next chapter gives detailed consideration to a number of electronically controlled systems so that the reader may assess the type of knowledge that is necessary for diagnosis, repair and maintenance of such systems.

Summary

In this chapter we have considered the need to revise and understand basic electrical principles and circuit diagrams. This aspect is important in dealing with electronic systems because much of the work in diagnosis and repair requires a person to test for faults in the circuits that connect the various elements of the electronic system together.

We have also considered some of the basic building blocks of electronics, e.g. p-n junctions, diodes, transistors and integrated circuits.

4
Exhaust emissions and engine management systems

Experience shows that people who have a good knowledge of vehicle systems and are able to carry out a good visual inspection of an electronic system before starting diagnostic work on it are normally able to eliminate 'obvious' faults without loss of time. In order to carry out this visual examination it is necessary to have an understanding of how the system operates.

This chapter takes a typical example of a vehicle electronic system as an indication of the type of knowledge that a technician needs to acquire about any system that he/she proposes to work on.

Atmospheric pollution

Atmospheric pollution is widely held to be a serious problem and in an attempt to reduce pollution of the atmosphere many countries have made laws that require vehicles to conform with certain levels of emissions. Figure 4.1 gives an indication of the types of atmospheric pollutants that cause most concern.

The regulations that govern vehicle emissions are based on the mass of pollutants that are emitted per mile (or kilometre) in certain driving conditions.

Fig. 4.1 shows the emission levels of the Environmental Protection Agency (EPA) of USA. They apply to diesel engines and the PM stands for particulate material, which includes carbon solids and metallic elements. The other levels refer to oxides of nitrogen, NO_X; carbon monoxide, CO; and hydro-carbons, HC. The last three are the ones of concern in the operation of petrol engines and I would now like to take a look at some electronic systems that have been developed to control these emissions.

In the UK once a vehicle has been type approved the emission checks are those that are carried out at the annual MOT, or during checks by approved Department of Transport officials. For petrol engined vehicles fitted with catalytic converters the gas analyser must show readings that conform to the manufacturer's specification. For diesel engined vehicles the opacity of the exhaust products is measured.

EPA Heavy-Duty Diesel Emission Regulations for Heavy-Duty Trucks & Urban Buses (g/bhp-hr)				
	NOx	HC	CO	PM
1988 (all)	10.7	1.3	15.5	0.60
1990 (all)	6.0	1.3	15.5	0.60
1991 (truck)	5.0	1.3	15.5	0.25
1991 (urban bus)	5.0	1.3	15.5	0.10
1994 (truck)	5.0	1.3	15.5	0.10

Fig 4.1 An example of exhaust emission limits

Figure 4.2 gives an indication of the processes that occur and which lead to 'pollution' of the atmosphere.

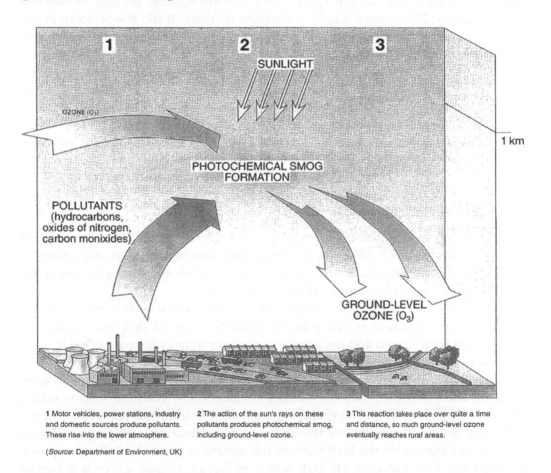

1 Motor vehicles, power stations, industry and domestic sources produce pollutants. These rise into the lower atmosphere.

2 The action of the sun's rays on these pollutants produces photochemical smog, including ground-level ozone.

3 This reaction takes place over quite a time and distance, so much ground-level ozone eventually reaches rural areas.

(*Source*: Department of Environment, UK)

Fig 4.2 Emissions and the atmosphere (D of E)

To meet emission regulations certain procedures have been developed, e.g. catalytic converters and exhaust gas recirculation. For these systems to function to the best advantage their operation needs to be co-ordinated with other engine systems, such as ignition timing and fuelling, and this is where electronic control comes in.

When a number of such systems are brought under the control of a single computer it is known as an engine management system (EMS).

After we have examined some individual systems we shall study a combined system, and then have a look at some of the servicing and diagnostic procedures.

NO_X

Motor fuels are hydro-carbons; that is to say they consist primarily of the two elements hydrogen and carbon. To enable the fuel to burn efficiently and release the energy that drives the engine, the fuel must be provided with an adequate supply of oxygen. The oxygen comes from the atmospheric air, and air contains approximately 23% oxygen and 77% nitrogen by mass. If the temperature in the combustion chamber rises above approximately 1800°C, some of the oxygen will react with the nitrogen to form nitrogen oxide, which is a harmful pollutant.

It thus seems fairly easy to say: 'Well, keep the temperature below 1800°C'. Unfortunately, it is not that simple. In order for the engine to operate efficiently it needs to have a high compression ratio and to achieve the highest possible combustion temperature. Obviously a compromise must be achieved and it seems that this may be done in one of several ways. The ones we shall examine here are: exhaust gas recirculation; retarding the ignition; and catalytic conversion.

EXHAUST GAS RECIRCULATION (EGR)

With exhaust gas recirculation some exhaust gas is drawn from the exhaust system and redirected back to the induction system. This has the effect of reducing the temperature in the combustion chamber and thus preventing the conditions of high temperature which produce NO_X.

There are times, during the normal driving cycle, when EGR is not useful, so permanent recirculation is not used. To make best use of EGR the quantity of exhaust gas recirculated and the phases of the operating cycle, when EGR is used, are placed under the control of a computer.

Figure 4.3 shows part of the TCCS (Toyota Computer Controlled System). Please focus on the EGR valve, the EGR vacuum modulator, VSV (vacuum sensing valve), and the ECU input to the VSV. The vacuum derives from the depression in the induction tract and the path of the recirculated exhaust gas may be traced from the exhaust system via the EGR valve to the induction tract. The ECU (computer) actuates the VSV and cuts off EGR when intake air volume exceeds a certain level.

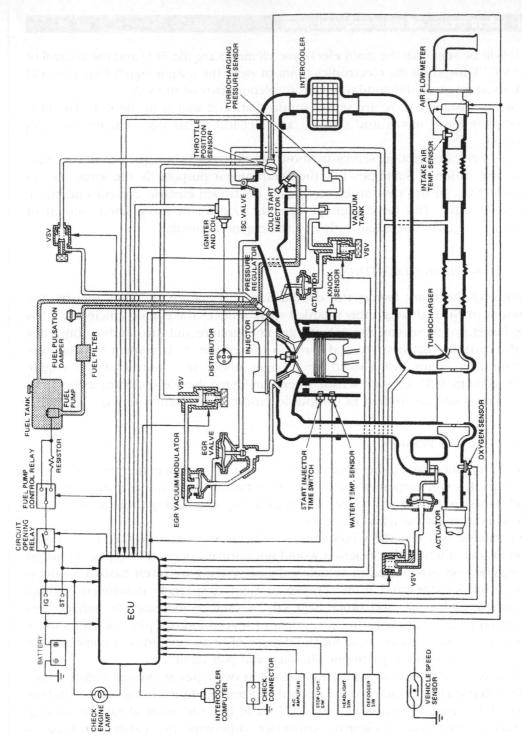

Fig 4.3 EGR section of Toyota TCCS Engine Management System

It will be seen that the main electronic elements are the ECU and the control of the VSV. Thus, from the electronics point of view, the output signal from the ECU is important, as is the condition of the solenoid part of the VSV.

In this instance, there are sensor inputs that send signals to the ECU. The ECU processes these inputs and outputs an electrical signal that initiates the EGR process.

This brief description serves to show one form of EGR: other engine makers may use different strategies but the fundamental purpose is the same, i.e. to reduce NO_x. In this case an electronically controlled electric current operates a solenoid valve. This valve then permits the transfer of a regulated amount of exhaust gas from the exhaust pipe to the engine air intake.

ALTERING THE IGNITION TIMING

Retarding the spark lowers the combustion temperature and thus reduces the amount of NO_x. Although this retarded spark reduces the engine efficiency, it has the benefit of increasing the exhaust gas temperature and this tends to burn any hydro-carbons that have passed through the engine. It is evident that ignition timing is a factor that should be brought under the control of the engine management ECU. A knock sensor, often employing the piezo electric effect, may be used as one controlling ECU input to ensure ideal ignition timing.

CATALYTIC CONVERSION

Ideally engines would be designed so as not to produce harmful pollutants. Perhaps this will happen at some time in the future, but as things stand at present pollutants are produced and the commonest way of reducing them to acceptable levels is to use a catalytic converter.

A catalyst is a material which causes a chemical reaction to take place under conditions in which that reaction would not normally occur. The types of catalyst (catalytic converter) used in vehicle exhaust systems are often based on honeycomb blocks fabricated from ceramic or steel. A specially developed coating, comprising aluminas and other materials, is used to increase the surface area exposed to exhaust gas to the equivalent of several football pitches. This surface coating provides a stable base for the deposit of small quantities of combinations of the precious metals platinum, rhodium and palladium.

The two types of catalyst formation result in two types of exhaust construction: the pelletised type which has the general appearance shown in Figure 4.6; and the monolithic type as shown in Figure 4.7. Lead free fuel must always be used in vehicles fitted with a catalytic converter, otherwise the catalyst material is 'poisoned' and ceases to function.

For catalytic converters to work effectively they must be provided with optimum conditions, such as operating temperature and correct air–fuel ratio of the engine fuelling. The general approach to this is to use electronically controlled systems because they seem to provide the best solution.

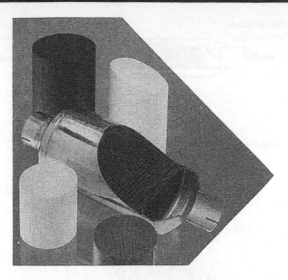

Fig 4.4 Honeycomb autocatalysts (Johnson Matthey)

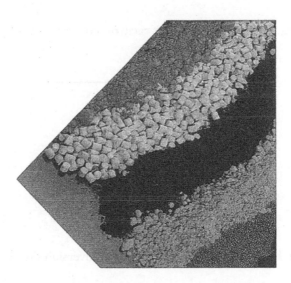

Fig 4.5 Pellet type catalyst (Johnson Matthey)

'Light off' temperature

Figure 4.8 shows how the conversion efficiency of an oxidising catalyst changes with temperature. There is a temperature, approximately 300°C, at which the catalyst functions effectively and this is known as the 'light off' temperature. It is evident that the conversion efficiency is low when the system is cold, and recent developments provide for preheating of the catalyst.

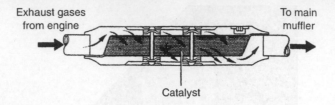

Exhaust gases
from engine

To main
muffler

Catalyst

Fig 4.6 Pellet type catalytic converter (Toyota)

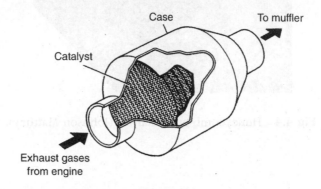

Case

To muffler

Catalyst

Exhaust gases
from engine

Fig 4.7 Monolithic type catalytic converter (Toyota)

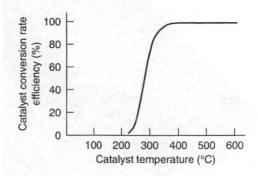

Fig 4.8 Catalyst temperature and conversion efficiency

Air-fuel ratio

It is usual to express air–fuel ratio in terms of mass of air in relation to mass of
fuel. A typical hydro-carbon fuel, such as petrol, has an air–fuel ratio for correct
(stoichiometric) combustion of approximately 14.7:1. That is 14.7 kg of air to
each 1 kg of fuel. Figure 4.9 shows the approximate relationship between air–fuel
ratio and catalyst conversion efficiency.

There is a small range (window) of air–fuel ratios where the catalyst conversion
efficiency is high and it is obviously advantageous to operate the engine in this
range.

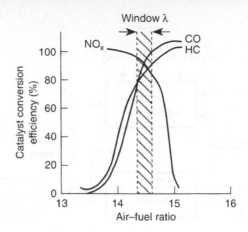

Fig. 4.9 Conversion efficiency of exhaust catalyst

The oxygen content of the exhaust gas provides a good guide to air–fuel ratio at the engine inlet and oxygen sensors have been developed which provide an input to an electronic system so that emission controls can be made to operate effectively.

The lambda sensor

Often the correct air–fuel ratio (stoichiometric) 14.7:1 is related to the Greek letter lambda in a way such that 14.7:1 = 1 (lambda).

If the air–fuel ratio is weak, i.e. more air, lambda is greater than 1, say 1.05; and if the air–fuel ratio is rich, i.e. less air, lambda is less than 1, say 0.95. When this nomenclature is used the oxygen sensor is called a lambda sensor. Otherwise it is known as the oxygen sensor.

Whatever the case the purpose is the same, which is to sample the oxygen content of the exhaust gas to provide a means of controlling the air–fuel ratio within the 'window', so that the conversion efficiency of the catalyst is maximised.

Figure 4.10 shows the position of the exhaust sensor and the other elements of an emission control system. By sampling the exhaust gas continuously and feeding back a voltage that represents oxygen levels in the exhaust, the ECU is able to alter the fuelling so that the catalyst can keep working.

Figure 4.11(a) shows a typical oxygen sensor. This type of sensor is mounted in the exhaust system close to the engine, as shown in Figure 4.11(b).

A much simplified description of the operation of the oxygen sensor is that it operates by permanently sampling the oxygen content of the atmospheric air and comparing it with the oxygen content of the exhaust.

CONSTRUCTION OF AN OXYGEN SENSOR

The probe is fitted in a housing which protects the ceramic body against mechanical damage. The outer part of the ceramic body is positioned in

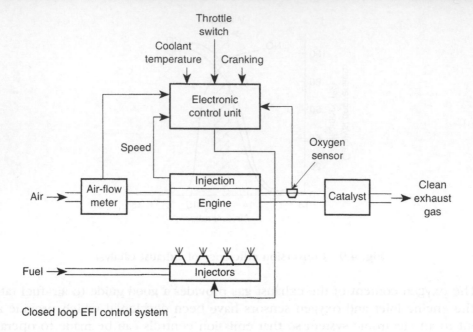

Closed loop EFI control system

Fig. 4.10 Emission control system (Lucas)

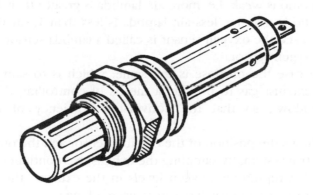

Fig. 4.11(a) Lucas type lambda (oxygen) sensor

the exhaust gas stream and the inner part is in contact with the ambient (atmospheric) air.

The ceramic body consists primarily of zirconium dioxide. Each surface (inner and outer) is coated with an electrode made of a thin layer of platinum which is permeable to gas. In addition, a porous ceramic layer is applied to the surface exposed to the exhaust gas. This layer protects the surface of the electrodes

Fig. 4.11(b)   The oxygen sensor in an exhaust pipe

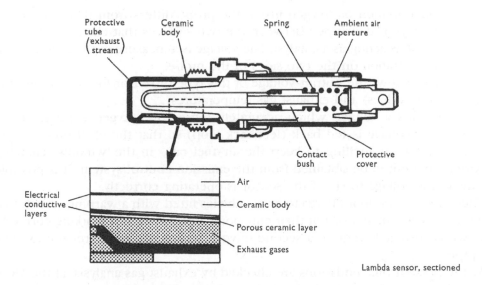

Lambda sensor, sectioned

Fig. 4.12   Details of Lucas type Lambda (oxygen) sensor

against contamination caused by combustion residues in the exhaust gas, thus ensuring that the operating characteristics of the sensor do not deteriorate. The ceramic material used becomes conductive for oxygen atoms at a temperature of approximately 300°C and above.

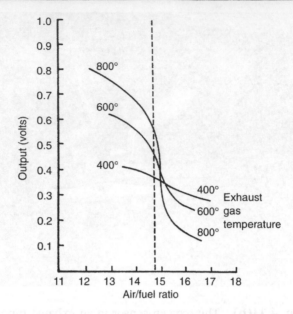

Fig. 4.13   Oxygen sensor voltage (Lucas)

If the concentration of oxygen inside the probe differs from that outside, an electrical voltage is developed between the two surfaces that changes when the outer electrode is acting as a catalyst. The voltage is a measure of the difference in oxygen concentration on the two sides of the probe.

Figure 4.13 shows the output voltage of this type of sensor for various air-fuel ratios (lambda values) and exhaust gas temperatures.

There is thus a voltage which accurately represents oxygen content of the exhaust. This voltage is fed back (feedback loop) so that the electronic control unit can adjust the fuelling to keep the air-fuel ratio in the 'window' enabling maximum benefit to be obtained from the emission control system. It is possible to check this voltage to see if the sensor is operating correctly.

There are moves afoot (1998) to have vehicles fitted with a warning system that will tell drivers whether or not their emission control system is working correctly. This would probably require a second oxygen sensor placed nearer the exhaust tail pipe.

At the present time emissions are checked by exhaust gas analysis, at the MOT, or possibly road-side checks. It should be understood that failure to meet emission standards does not necessarily mean that the catalytic converter has failed. Any defect that causes the combustion to be incorrect could lead to incorrect emissions and it is imperative to ensure correct operation of the system before the drastic step of replacing the entire exhaust system is contemplated.

This provides a convenient point at which to switch to an engine management system so that we can discuss some of the factors that have to be considered in maintenance and repair.

An engine management system (EMS)

Engine management is a term which is used in a fairly general way to describe an electronically controlled engine system. The degree of complexity of an EMS varies from make to make of vehicle. For our purposes it is sufficient to consider a fuel system. From this we can gain an idea of the working of the system and the types of test that can usefully be carried out to ascertain the condition of various components in the system. The procedures established can then, with the aid of information about a specific system, be extended to cover a range of systems.

Figure 4.14 shows a fairly typical electronic fuel injection (EFI) system. I have already described how the oxygen sensor provides a computer input which is used to help control emissions. This EFI does not incorporate a catalytic exhaust but it does have several features in common with other, more complicated, systems. It is these common features that I now intend to focus on.

Examination of Figure 4.14 will show that there are several sensors providing computer inputs and several actuators that respond to computer outputs. All

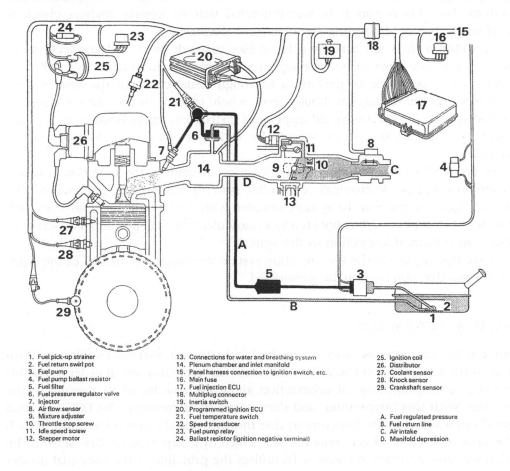

1. Fuel pick-up strainer	13. Connections for water and breathing system	25. Ignition coil
2. Fuel return swirl pot	14. Plenum chamber and inlet manifold	26. Distributor
3. Fuel pump	15. Panel harness connection to ignition switch, etc.	27. Coolant sensor
4. Fuel pump ballast resistor	16. Main fuse	28. Knock sensor
5. Fuel filter	17. Fuel injection ECU	29. Crankshaft sensor
6. Fuel pressure regulator valve	18. Multiplug connector	
7. Injector	19. Inertia switch	
8. Air flow sensor	20. Programmed ignition ECU	
9. Mixture adjuster	21. Fuel temperature switch	A. Fuel regulated pressure
10. Throttle stop screw	22. Speed transducer	B. Fuel return line
11. Idle speed screw	23. Fuel pump relay	C. Air intake
12. Stepper motor	24. Ballast resistor (ignition negative terminal)	D. Manifold depression

Fig. 4.14 Electronic fuel injection

these devices need to be working properly if the entire system is to operate efficiently.

Examples of sensors included in the system are:

- manifold pressure
- throttle position
- coolant temperature

Examples of actuators are:

- the main fuel injector
- the air valve actuator
- the cold start injector

Before examining the detail of these devices it is well to remember that electronic devices are very reliable. If there *is* a system defect it is more likely that it is due to some ordinary motor vehicle problem, such as a dirty air filter, a fouled spark plug or some other fault associated with the non-electronic parts of the system. For this reason it is recommended that an engine analysis check is performed before attempting any electronic diagnosis.

In addition, it is not unknown for cables to work loose and connections to sensors and actuators to become defective. To reduce the prospect of defective analysis it is advisable to perform a thorough visual check of the system. This requires that the technician should know where to look for the various elements of the system. To aid this visual inspection instruction manuals normally include location diagrams that show where the various elements are located on the vehicle.

Figure 4.15 shows an example of a location diagram of the type that is useful when working on electronically controlled systems. It should be appreciated that a working knowledge of the system, coupled with the knowledge of what should, and what should not, be connected to a particular unit, will permit an observer to make an informed inspection of the system.

If we now return to the fuel injection system we should get an idea of the types of checks that can usefully be performed.

AIR FLOW MEASUREMENT

An engine requires the correct air–fuel ratio to suit various conditions. With electronic fuel injection the ECU controls the air–fuel ratio and in order to do this it needs a constant flow of information about the amount of air flowing to the engine. With this information, and data stored in its memory, the ECU can then send out a signal to the injectors so that they provide the correct amount of fuel. Air flow measurement is commonly performed by a 'flap' type air-flow sensor. The air-flow sensor shown in Figure 4.16 utilises the principle of the potential divider (potentiometer).

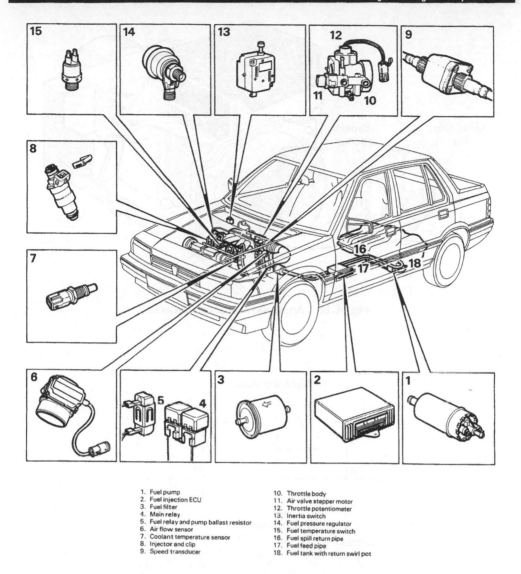

1. Fuel pump
2. Fuel injection ECU
3. Fuel filter
4. Main relay
5. Fuel relay and pump ballast resistor
6. Air flow sensor
7. Coolant temperature sensor
8. Injector and clip
9. Speed transducer
10. Throttle body
11. Air valve stepper motor
12. Throttle potentiometer
13. Inertia switch
14. Fuel pressure regulator
15. Fuel temperature switch
16. Fuel spill return pipe
17. Fuel feed pipe
18. Fuel tank with return swirl pot

Fig. 4.15 Location diagram for an electronic system

Figure 4.17 shows the theoretical form of a simple potential divider. A voltage, say 5 V, is applied across terminals A and B. C is a slider which is in contact with the resistor and a voltmeter is connected between A and C. The voltage V_{AC} is related to the position of the slider C in the form $V_{AC} = V_{AB} \times X/1$.

In the air-flow sensor the moving probe (wiper) of the potential divider is linked to the pivot of the measuring flap so that angular displacement of the measuring flap is registered as a known voltage at the potentiometer.

Figure 4.18 shows a simplified form of the air-flow sensor. The closed position of the measuring flap will give a voltage of approximately zero, and fully open the voltage will be 5 V. Intermediate positions will give voltages between these values.

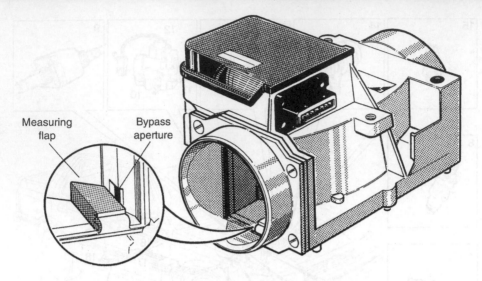

Measuring flap

Bypass aperture

Fig. 4.16 An air-flow sensor (Lucas)

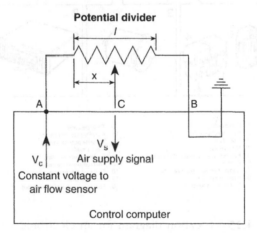

Potential divider

A

C

B

V_c

V_s

Air supply signal

Constant voltage to air flow sensor

Control computer

Fig. 4.17 A simple potential divider

In practice, it is not quite as simple as this because allowance must be made for other contingencies. However, this should not detract from the value of knowing the basic principles because it is these which lead to the diagnostic checks that can be applied. Before we consider some basic checks that can be performed we should have a more detailed look at the Lucas air-flow meter of Figure 4.16.

Figure 4.19 shows the principle of the Lucas 2AM sensor. The relative position of the measuring flap depends on the air flowing to the engine and the return torque of the spring. This diagram gives a little more information because it shows

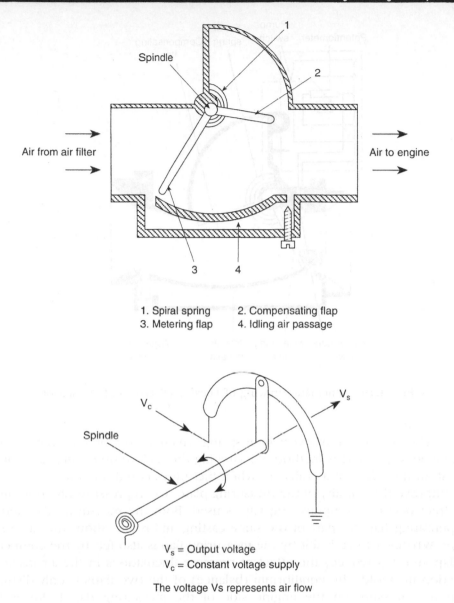

1. Spiral spring   2. Compensating flap
3. Metering flap   4. Idling air passage

V_s = Output voltage
V_c = Constant voltage supply

The voltage Vs represents air flow

Fig. 4.18 The potential divider applied to an air-flow sensor

a refinement – the idle air by-pass – that is required for satisfactory operation of the device.

To compensate for production tolerances and different air requirements for similar engines at idling speed it is desirable that air–fuel ratio can be adjusted under idling conditions. The idle air by-pass together with the adjustment screw permits the idle mixture to be adjusted to suit individual engine requirements. The air-flow meter (sensor) incorporates a fuel pump switch. This switch is controlled by the initial (approximately 5°) movement of the measuring flap. This

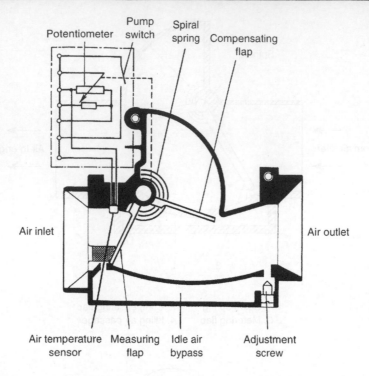

Potentiometer Pump switch Spiral spring Compensating flap

Air inlet   Air outlet

Air temperature sensor   Measuring flap   Idle air bypass   Adjustment screw

Fig. 4.19 Schematic drawing of Lucas 2AM type air-flow sensor

'free play' is controlled by a very light spring which is overcome by very small air flow, i.e. idle speed. This method of switching the fuel pump ensures a minimum fire risk in the event of a collision, when fuel pipes could fracture.

To improve the stability of the measuring plate, when pressure variations in the inlet tract occur, a compensating flap is used. Both the measuring flap and the compensating flap are part of the same casting and rotate about the same shaft centre. Whatever force is felt by the measuring flap is also felt by the compensating flap and this reduces the effect of pressure fluctuations in the air inside the induction manifold. The equilibrium (balance) of the two flaps is only disturbed when the pressure on the engine side of the measuring flap is lower than atmospheric. This pressure is also felt on both sides of the compensating flap.

AIR-TEMPERATURE SENSOR

The air-flow sensor also incorporates a temperature sensor and this also sends a signal to the ECU which uses it to calculate the mass of air. (Mass and volume of air are linked by temperature.)

Testing the air-flow sensor
It should be noted that the kind of air-flow sensor covered here is one of many types. The tests described relate only to this particular type. However, these tests

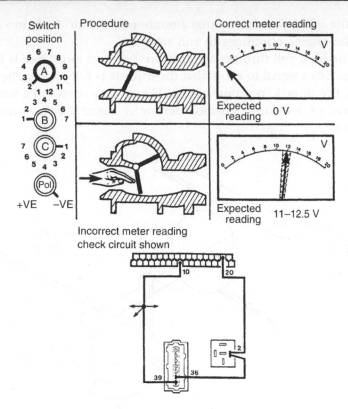

Fig. 4.20   Testing an air-flow sensor

do illustrate, in broad terms, the type of tests that can be performed on electronic systems. Similar procedures can be applied to other types of air-flow sensors, but it would first be necessary to study the working principles and gain access to the necessary test data.

Figure 4.20 shows a voltmeter connected to the appropriate terminals and the meter flap being moved by hand. The voltage reading will vary as the flap is moved from closed to open position. The test shown here is indicative of the type of simple tests that can be performed on some components of an electronically controlled system. However, it should be understood that full knowledge of the vehicle being worked on must be to hand; this type of test would not work on a 'hot-wire' type air-flow meter. The type of specific information about a particular vehicle that will be required for test purposes includes wiring diagrams, connector pin identification numbers and other important data. These details are only normally available in the manufacturer's workshop manuals.

THROTTLE POSITION

When an engine is idling the exhaust gas scavenging of the cylinders is poor. This has the effect of diluting the incoming mixture. The ECU must detect when the throttle is in the idling position so that alteration of the air-fuel ratio can occur to ensure that the engine continues to run smoothly.

At full engine load, full throttle, the mixture (air-fuel ratio) needs enriching, so the ECU also needs a signal to show that the throttle is fully open. These duties are performed by the throttle position switch. Figure 4.21 shows in a simplified form how the action of a throttle position sensor is based on the principle of the potential divider.

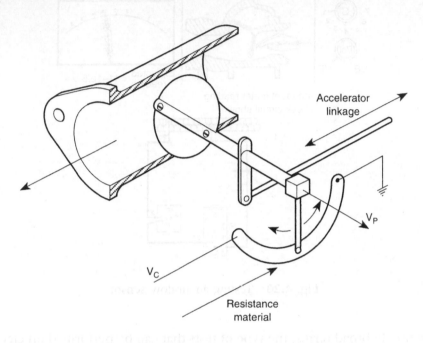

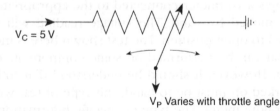

V_C = Constant voltage supply from computer

V_P = Voltage giving position of throttle

Fig. 4.21 The principle of the throttle position sensor

The sensor produces a voltage which is related to throttle position, and the voltage signal is conducted to the ECU where it is used, in conjunction with other inputs, to determine the correct fuelling for a given condition.

There are two types of throttle position sensors in common use; they are quite different in certain respects, and test procedures that will work on one type will

produce, and which are used by the electronic

Contact type

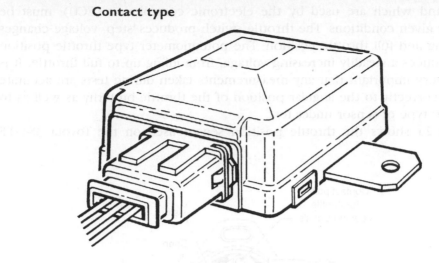

Fig. 4.22 Lucas type throttle position switch

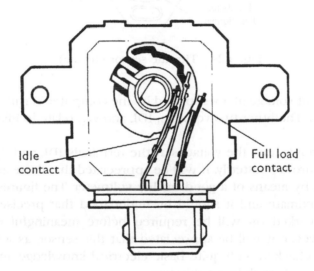

Idle contact

Full load contact

Fig. 4.23 Inside the throttle switch (Lucas)

not necessarily apply to the other type. Here again it is important to be able to recognise which type is being used in a particular application.

Figures 4.22 and 4.23 show details of the throttle position switch.

The point to note about these two types of throttle position sensors is that they are primarily electrical. They do not require any great electrical or electronic knowledge in order to test them; however, the electrical signals which they produce, and which are used by the electronic control unit (ECU), must be correct for given conditions. The throttle switch produces 'step' voltage changes at the idling and full throttle position. The potentiometer type throttle position sensor produces a steadily increasing voltage, from idling up to full throttle. It is therefore very important that any measurements taken during tests are accurate and relate correctly to the angular position of the throttle butterfly as well as to the specific type of sensor under test.

Figure 4.24 shows the throttle position sensor used on the Toyota 3S-GTE engine.

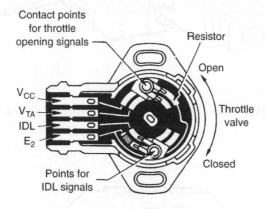

Fig. 4.24 Throttle sensor (Toyota)

V_{CC} is a constant voltage of 5 V supplied by the computer. Terminal E_2 is earthed via the computer. The other two voltages, IDL and V_{TA}, relate to idling and throttle operating angle.

Figure 4.25 shows how the voltages at the terminals IDL and V_{TA} relate to the position of the throttle butterfly. It will be appreciated that these are voltages that can be checked by means of a good quality voltmeter. The figures shown in the graph are approximate and it should be understood that precise details of any vehicle being worked on will be required before meaningful checks can be performed. However, it will be appreciated that this sensor, as with many other sensors, can be checked with quite basic electrical knowledge and good quality widely available tools such as a voltmeter.

Let us recap. The ECU generally requires special tools to verify its condition. The ECU is but one part of an electronic system. All other elements of the system must be working properly in order for the system, overall, to function correctly. All sensors and actuators must function accurately, and all interconnecting circuits must be complete. Much of the system, external to the ECU, consists of conventional technology and can be checked without a great deal of electronic expertise.

So far in this chapter we have been looking at some of the types of checks that can be performed with the aid of multi-meters, and information about the system.

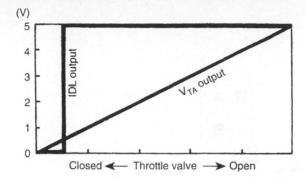

Fig. 4.25 Indication of throttle sensor voltages (Toyota)

Before moving on to consider other tests, it is well just to remind ourselves about some of the computer inputs that are used in a fuel injection system.

Figure 4.26 shows the inputs to a fuel injection system ECU. Up to this point we have seen that air-flow, throttle position and oxygen sensors each produce voltage outputs that can be measured. We will now look at the other inputs and the fuel injection duration output, for it would seem that if we can evaluate all ECU inputs and outputs we should be well on the way to evaluating the performance of the whole system, particularly if we are able to check the interconnecting circuits.

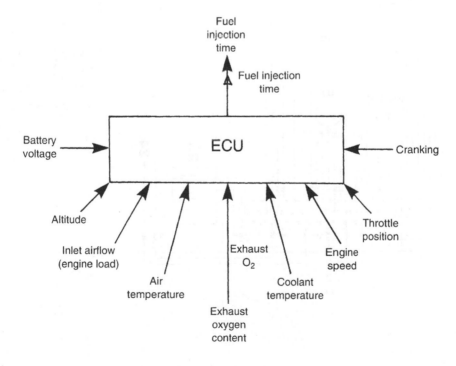

Fig. 4.26 Computer inputs (Lucas)

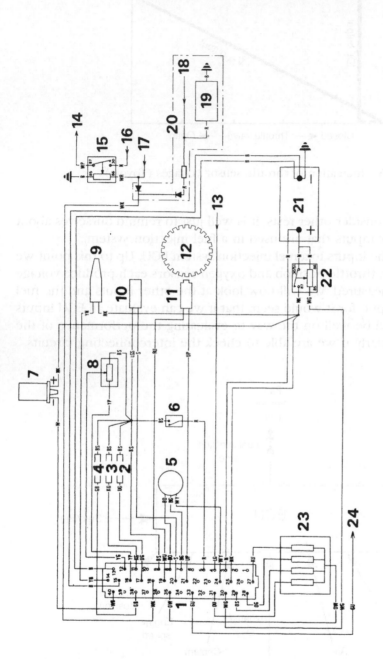

1. 40-way harness connector
2. Ambient air sensor
3. Coolant sensor
4. Inlet air sensor
5. Diagnostic connector
6. Throttle pedal switch
7. Ignition coil
8. Throttle potentiometer
9. Injector
10. Knock sensor
11. Crankshaft sensor
12. Reluctor disc

13. Feed from ignition switch
14. To starter relay
15. Starter relay
16. Positive feed from battery
17. Cranking input (ignition switch)
18. Feed from air-con. relay
19. Air-con. magnetic clutch
20. Air-con. fitment only
21. Battery
22. ECU relay
23. Stepper motor
24. To temperature gauge

Fig. 4.27 Crank speed and position sensor. 11. Crankshaft sensor; 12. Reluctor disc

ENGINE SPEED

It is generally reckoned that the crank gives a good reference point for detecting engine speed, so we shall have a look at an engine speed sensor that uses the crankshaft as a reference. The principle of other engine speed sensors is similar even though they may use the camshaft or ignition distributor as a reference.

Figure 4.27 shows a typical crankshaft speed sensor. The steel reluctor disc is fitted to, and rotates with, the engine flywheel. The sensor is mounted on the engine casting and is located so that the passage of the reluctor disc segments, past the sensor armature, generates an electric current in the sensor winding. Figure 4.28 shows the location of the sensor and its wiring harness.

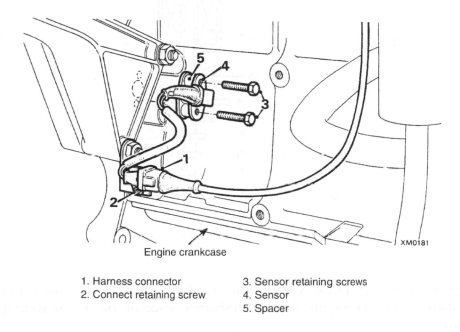

Engine crankcase

1. Harness connector
2. Connect retaining screw
3. Sensor retaining screws
4. Sensor
5. Spacer

Fig 4.28 Crankshaft sensor in position on the engine

Figure 4.29 shows how the sensor is constructed. It will be seen that it has a permanent magnet, a coil and some terminals, A & B, through which the electrical signals are passed to the ECU.

Because the current induced in the winding is dependent on the speed of approach and departure of each reluctor segment, in relation to the sensor armature, the voltage is of an alternating form, as shown in Figure 4.30.

As the operation of this sensor is not dependent on an electrical input it should be evident that rotation of the crankshaft will give rise to an electrical output, at the terminals A & B. It should, therefore, be possible to check that the sensor is functioning correctly by connecting a voltmeter to those terminals, when disconnected from the harness, as shown in 4.29(a), and then rotating the crankshaft by

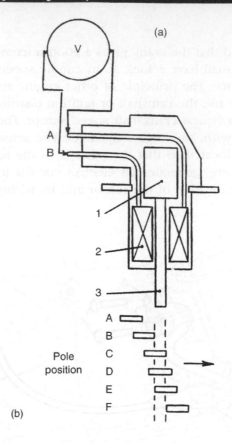

Fig 4.29   Construction of the crank speed sensor

means of the starter motor. A static test can also be applied with the aid of an ohm-meter. This will require that the resistance value of the sensor winding is known.

The voltage signal from the crank sensor is electronically processed to make it suitable for use in the control unit (to indicate crank position and engine speed) and then it is used to control fuel injection point and ignition timing.

It should now be clear that many of the sensors rely for their operation on conventional electrical principles and that they may be checked for correctness of operation by the use of a good quality multi-meter. Before moving on to assess what can be done to test actuators, I think we should take a look at one more sensor because I believe that we shall then have acquired sufficient understanding to have established some useful ideas about sensor inputs to the ECU.

COOLANT TEMPERATURE SENSOR

In Chapter 2, Figure 2.2 shows a typical coolant temperature sensor, and Figure 2.4 shows a table of resistance values for given temperatures. From this it may be

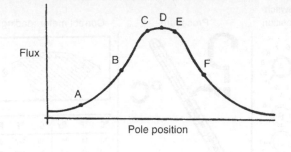

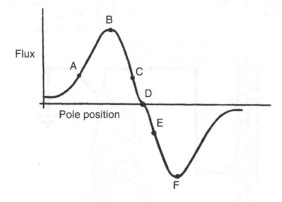

Fig 4.30 Crank sensor voltage waveform

seen that a resistance and coolant temperature test will show whether or not the sensor is working correctly.

Figure 4.31 shows the settings for some specialist equipment that was developed for electronic systems testing. This could be replaced by a thermometer and an ohm-meter so that resistance may be checked against temperature. The readings obtained may then be compared with a table, such as the one in Figure 2.4. The lower part of Figure 4.31 shows that the entire circuit between the ECU, terminal 13 and the sensor, and between the sensor and earth, should be investigated in the event of an incorrect resistance versus temperature check.

INERTIA SWITCH

Before concluding this section, it must be stated that petrol fuel systems operate at quite high pressures and work on the fuel lines, etc. should only be performed by competent persons.

In order to reduce fire risk, in the event of a collision, the fuel system is fitted with a fuel pump cut-off switch. This switch, of the type shown in Figure 4.32, is inertia sensitive and for this reason it is often referred to as an inertia switch.

In the event of a fuel system failure the inertia switch should be checked.

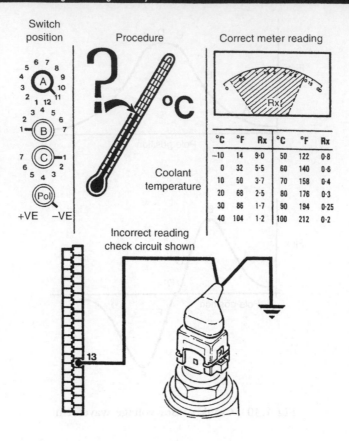

Switch position	Procedure	Correct meter reading

Coolant temperature

°C	°F	Rx	°C	°F	Rx
−10	14	9·0	50	122	0·8
0	32	5·5	60	140	0·6
10	50	3·7	70	158	0·4
20	68	2·5	80	176	0·3
30	86	1·7	90	194	0·25
40	104	1·2	100	212	0·2

Incorrect reading
check circuit shown

Fig 4.31

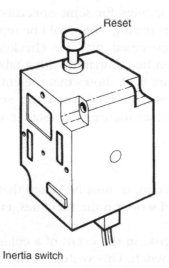

Fig 4.32 Inertia switch

Summary

In this chapter we have considered some of the environmental factors that have led to the use of electronically controlled systems. It should be noted that improved engine efficiency is also achieved by the use of electronic systems. We have seen that the ECU requires inputs from sensors, it then processes them and sends outputs to actuators. We have seen that many sensors operate on conventional electrical principles and this allows comprehensive testing to be performed with the aid of a multi-meter – and a sound knowledge of basic electrical principles.

The range of sensors covered is indicative of the type of technology involved, but it should be understood that there are very many more types of sensors used on vehicles. This chapter merely covers sufficient to give an understanding of the principles involved and the types of tests that can be performed. Other circuit tests are covered in Chapter 6.

In a book such as this it is not appropriate to consider detailed testing because vehicle systems vary greatly. You have only to consider the throttle position potentiometer and the throttle position switch. With the potentiometer a constantly varying resistance is to be expected, between throttle closed and throttle fully open, whereas with the throttle switch there will be step changes. This is why a knowledge of basic principles is so important; once these are understood it is necessary to have accurate information about the specific vehicle that is being worked on. However, once the data has been acquired it is evident that a great deal of useful testing can be performed with the aid of fairly basic equipment such as a multi-meter for measuring volts, amperes and ohms, and a carefully developed strategy, such as the 'six step' approach.

ACTUATORS AND ECU OUTPUTS

The point has been made, and repeated, that electronically controlled vehicle systems have a feature in common, which is that they comprise sensors (inputs), actuators (operated by outputs), an ECU and interconnecting cables.

To open this chapter the relation between emission control and engine management was discussed. Some of the factors relating to emission control were covered and then detailed coverage of a fuel injection system was undertaken. This led to consideration of the types of sensors that are used to provide controlling inputs to the ECU. These inputs are processed by the ECU to give controlling outputs so that fuel injectors operate correctly and engine performance and emissions are as required by the designer. Should the system not work correctly and power is low, or emissions are out of limits, then it may be necessary to check that the output (actuators) are working correctly.

INJECTORS

The commonest actuators on a petrol injection system are the injectors. Two types of petrol injection are used. One is single point, or throttle body injection.

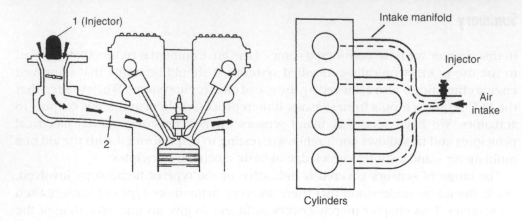

Fig 4.33   Single-point injection

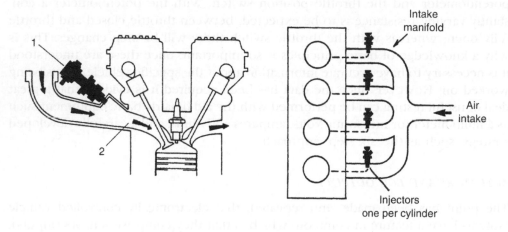

Fig 4.34   Multi-point injection

Here, in Figure 4.33, a single injector sprays fuel into the main air intake near the throttle butterfly. The other, in Figure 4.34, is multi-point where a separate injector is placed near the inlet valve for each cylinder.

As described in Chapter 3, the petrol injector relies for its operation on the principle of the electro-magnetic solenoid. Figure 4.35 shows the construction of a typical petrol injector.

The amount of fuel delivered by each injector is determined by the period of time for which the injector is kept open. The period varies from approximately 1.5 to 10 milliseconds, and this time is controlled by the ECU which, in response to sensor inputs and comparisons made with data stored in memory, will send out electrical pulses of predetermined length, between approximately 1.5 and 10 milliseconds.

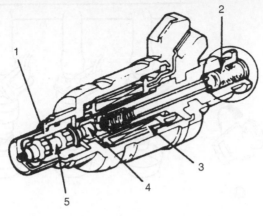

1. Plunger housing   4. Core
2. Filter   5. Plunger
3. Solenoid coil

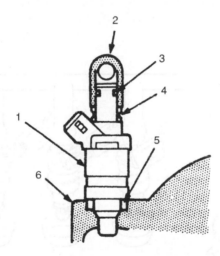

1. Injector   4. Cushion ring
2. Fuel pipe   5. Seal ring
3. 'O' ring   6. Intake manifold

Fig 4.35 Petrol injector

Various manufacturers make injector testing equipment and Figures 4.36 and 4.37 show examples of two tests that are possible on a vehicle. Again, these should only be undertaken by competent personnel, in possession of the necessary instructions and who are careful to take all required safety precautions.

Figure 4.36 illustrates the type of test that can be performed to check for injector leakage. The injectors are removed from the manifold and securely placed over drip trays. Note that they remain connected to the fuel and electrical system.

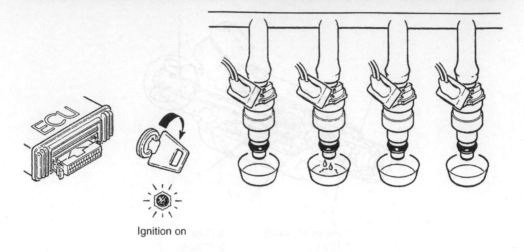

Ignition on

Fig 4.36 Testing injectors for leakage

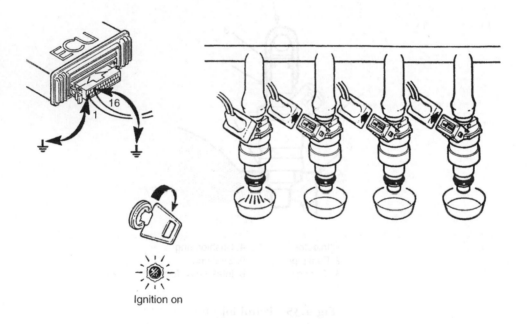

Ignition on

Fig 4.37 Testing the amount of fuel per injector

The approved procedure is then applied to energise the injectors and a careful visual inspection will reveal if there are any leaks. In the case shown it is recommended that any injector that leaks more than two drops of fuel per minute should be replaced.

Figure 4.37 shows the same injectors. In this case three of them are disconnected and the amount of fuel delivered by the injector which is still connected is carefully measured. It is possible to obtain special injector test benches on which these tests can be performed.

ELECTRICAL CONDITION OF INJECTORS

Figure 4.35 shows that an injector is essentially a coil of wire through which electric current flows to create a magnetic field. The coil has a known electrical resistance which may be checked by disconnecting the leads from the injector and connecting an ohm-meter to the two terminals. Figure 4.38 shows a wiring (circuit) diagram for the injectors of a fuel system. The units numbered 1 to 6 are the injectors.

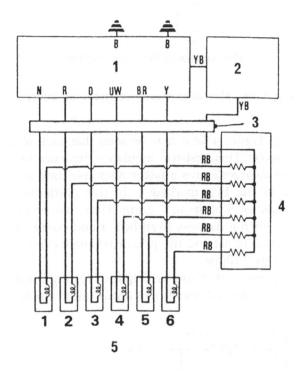

1. ECU
2. Main relay
3. Harness block conector R.H. suspension tower
4. Resistor pack
5. Injectors

Fig 4.38 Fuel injector circuit

Figure 4.39 shows one of the injectors being tested for resistance of the coil winding. The ohm-meter must be set to the correct range so that the resistance value of 1.5 to 2.5 ohms may be checked accurately. This resistance value will vary according to the type of injector being tested. In the case shown, the injector should be replaced if the resistance falls outside the range of 1.5 to 2.5 ohms.

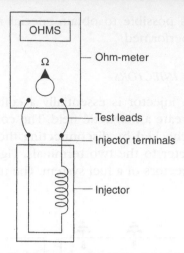

Resistance should be 1.5 to 2.5 ohms

Fig 4.39 Electrical resistance check on injector

By referring back to Figure 4.38 it may be seen that each injector is connected in series with a resistor. This resistor regulates the current that flows in the injector winding and also generates a faster response time at the injector. These resistors are shown in the box marked 4. In this particular case the resistors marked RB have a value of 5 ohms to 7 ohms. They may be checked individually by disconnecting the wiring harness from box 4 and connecting the ohm-meter between each resistor terminal and the power terminal (a). The reading obtained should be between 5 to 7 ohms.

These details are given so that the principle of testing output devices may be appreciated. It will be seen that the procedure is quite simple. However, it must

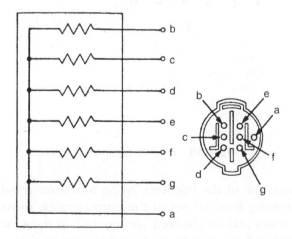

Fig 4.40 Testing the resistor block 4 (Rover)

be emphasised that such tests require the operative to be skilled and in possession of all the information that relates to a particular type of vehicle.

Safety note

Sparks, hot engines and petrol are a dangerous combination and the importance of always following safe procedures cannot be overstated. Among other factors, work on fuel systems should never be attempted in a confined space.

In the cases discussed in this chapter, it is evident that the main skill and knowledge required is based on electrical and motor vehicle principles and not on electronics. The bulk of the electronics is in the ECU. It should also be evident that knowledge of basic electricity is fundamental to developing the skill necessary for succesful testing of electronically controlled vehicle systems.

Methodical, safe and accurate working practices, based on sound knowledge of fundamental principles and correct use of instruments, together with information that relates to the system being worked on, are essential ingredients of successful workmanship on vehicle electronic systems.

5
Anti-lock braking systems (ABS) and cruise control

I have previously stated that most vehicle electronic systems have a feature in common, and that is that they consist of an ECU, actuators, sensors and circuits. I consider it useful therefore to examine a couple of other systems in order to justify this belief.

Anti-lock braking is another commonly used system, which we shall now examine with a view to establishing the commonality between electronically controlled vehicle systems.

The term ABS covers a range of electronically controlled systems that are designed to provide optimum braking in difficult conditions. ABS systems are used on many cars, commercial vehicles and trailers. In this chapter the intention is to examine some features of ABS and to show that here, as with engine management and other systems, many of the sensor inputs to the ECU can be verified by the use of ordinary instruments. Interconnecting circuits can also be checked, and the actuators often rely on conventional electrical principles for their operation, which usually means that they are amenable to some reasonably straightforward checks. Here, as with other systems, most of the electronics are in the ECU.

The purpose of anti-skid braking systems is to provide safer vehicle handling in difficult conditions. If wheels are skidding it is not possible to steer the vehicle correctly and a tyre that is still rolling, not sliding, on the surface, will provide a better braking performance. ABS does not usually operate under normal braking. It comes into play in poor road surface conditions - ice, snow, water, etc. - or during emergency stops.

Figure 5.1 shows a simplified diagram of an ABS system that gives an insight into the way such systems operate. The master cylinder (1) is operated via the brake pedal. During normal braking manually developed hydraulic pressure operates the brakes, and should an ABS defect develop the system reverts to normal pedal operated braking. The solenoid operated shuttle valve (2) contains two valves, A and B. When the wheel sensor (5) signals to the ABS computer (ECU) (7) that driving conditions require ABS control, a procedure is initiated which energises

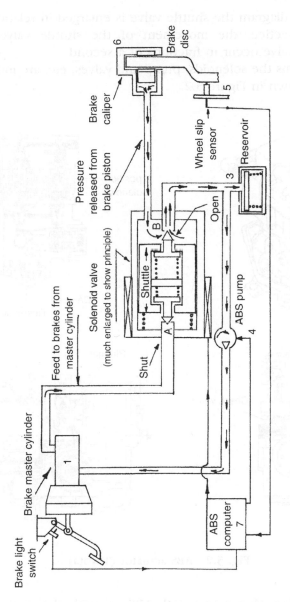

Fig. 5.1 A simplified version of an ABS

the shuttle valve solenoid. The valve (A) blocks off the fluid inlet from the master cylinder and the valve (B) opens to release brake line pressure at the wheel cylinder (6) into the reservoir (3) and the pump (4) where it is returned to the master cylinder.

In this simplified diagram the shuttle valve is enlarged in relation to the other components. In practice, the movement of the shuttle valve is small and movements of the valve occur in fractions of a second.

In practical systems the solenoids, pump and valves, etc. are incorporated into a single unit, as shown in Figure 5.2.

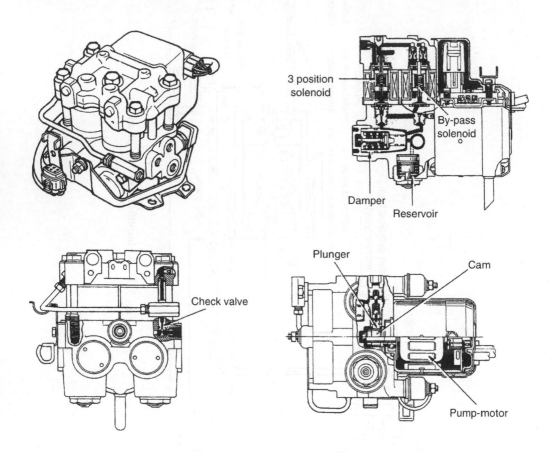

Fig. 5.2 ABS actuator (Toyota)

This brief overview shows that with ABS, as with the engine management system, the basic structure of the system is similar. That is to say, there are sensors, an actuator, an ECU and interconnecting circuits. To enable the whole system to function properly each of the separate elements needs to be working correctly.

In deciding whether or not a vehicle wheel is skidding, or on the point of doing so, it is necessary to compare the rotational movement of the wheel and brake disc, or drum, with some part such as the brake back plate which is fixed to the vehicle. This task is performed by the wheel speed sensing system. The procedure for doing this is reasonably similar in all ABS systems, so the wheel speed sensor is a good point at which to delve a little deeper into the operating principles of ABS.

The wheel speed sensor

Figure 5.3 shows a typical wheel sensor and reluctor ring installation. The sensor contains a coil and a permanent magnet. The reluctor ring has teeth and when the ring rotates past the sensor pick-up the lines of magnetic force in the sensor coil vary. This variation of magnetic force causes a varying voltage (emf) to be induced in the coil and it is this varying voltage that is used as the basic signal for the wheel sensor. This particular application is for a Toyota, but its principle of operation is typical of most ABS wheel speed sensors.

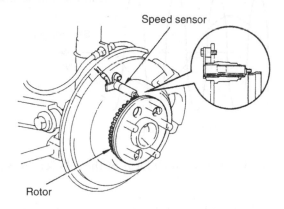

Fig. 5.3 ABS wheel sensor (Toyota)

The raw output voltage waveform from the sensor is approximately of the form shown in Figure 5.5. It will be seen that the voltage and frequency increase as the wheel speed, relative to the brake back plate, increases. This property means that the sensor output is a good representation of the wheel behaviour relative to the back plate and thus to provide a signal that indicates whether or not the wheel is about to skid.

In most cases this raw curved waveform is not used direct in the controlling process and it has first to be shaped, to a rectangular waveform, and tidied up before being encoded for control purposes. This conversion is achieved by means of integrated circuits and other devices, which are described in Chapter 8. The raw signal may be conveyed to the ECU for conversion at the analogue to digital interface, or it may be processed to rectangular form in a circuit built into the sensor unit.

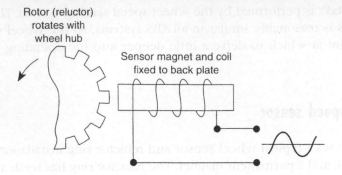

Rotor (reluctor)
rotates with
wheel hub

Sensor magnet and coil
fixed to back plate

Fig. 5.4   The basic principle of the ABS sensor

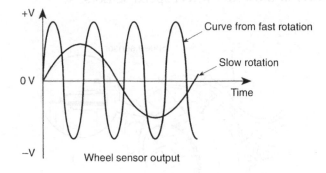

+V

Curve from fast rotation

Slow rotation

0 V

Time

−V

Wheel sensor output

Fig. 5.5   The voltage waveform of the ABS sensor output

If the brake is applied and the reluctor (rotor) starts to decelerate rapidly, relative to the sensor pick-up, it is an indication that the wheel rotation is slowing down. If the road surface is dry and the tyre is gripping well the retardation of the wheel will match that of the vehicle and normal braking will occur. However, if the road surface is slippery, a sudden braking application will cause the reluctor rotor, and road wheel, to decelerate at a greater rate than the vehicle, indicating that a skid is about to happen. This condition is interpreted by the electronic control unit. Comparisons are made with the signals from the other wheel sensors and the brake line pressure will be released automatically for sufficient time (a fraction of a second) to prevent the wheel from locking.

In hydraulic brakes on cars the pressure release and re-application is achieved by solenoid valves, a pump and a hydraulic accumulator, and these are normally incorporated into one unit called the actuator. The frequency of 'pulsing' of the brakes is a few times per second, depending on conditions, and the pressure pulsations can normally be felt at the brake pedal.

With air brakes on heavy vehicles the principle is much the same except that the pressure is derived from the air braking system and the actuator is called a modulator. The valves that release the brakes during anti-lock operation are solenoid operated on the basis of ECU signals, and the wheel sensors operate on the same principle as those on cars.

As for the strategy that is deployed to determine when to initiate ABS operation, there appears to be some debate. Some systems operate what is known as select low, which means that brake release is initiated by the signal from the wheel with the least grip, irrespective of what the grip is at other wheels. An alternative strategy is to use individual wheel control. Whichever strategy is deployed the aim is to provide better vehicle control in difficult driving conditions, and it may be that the stopping distance is greater than it would be with expert manual braking.

A commercial vehicle ABS

Throughout this book it has been my intention to show that electronically controlled vehicle systems have certain features in common. This is an important point to grasp because it provides a basis for the study of such systems, and also a basis for the methodical approach to maintenance and repair. In keeping with this philosophy I think it is useful to consider an anti-lock braking system that is used in conjunction with compressed air braking on large vehicles.

As with the motor car system, the basic idea is to prevent the wheels locking up under severe braking conditions, so that better control of the vehicle is achieved.

The term 'slip ratio' is used to quantify the percentage by which the wheel speed is less than the vehicle speed. It is generally believed that optimum braking performance in difficult conditions is achieved when the slip ratio is in the 15% to 30% range. The fundamental 'grip' between the surface and the tyre is called the coefficient of adhesion (coefficient of friction). Figure 5.6 shows how this coefficient varies with slip ratio for different surfaces. A map (computer program) that represents this graph is stored in the ECU memory so that it can be used as a reference. The ECU processor compares wheel/vehicle speed sensor information with the information in the memory, and if the slip ratio drifts beyond the ECU program's predetermined limits an electrical signal is sent to a solenoid that opens an air valve that releases air pressure from the brake on the wheel that is about to lock. When the wheel speed sensor signals that the slip ratio has come back within the reference range the ECU processor closes the air pressure bleed valve and opens another valve that restores braking power.

Because the ECU processor acts at high speed it is able to cope with rapidly changing braking conditions. For example, if there is ice under one wheel and dry tarmac under the wheel on the other side of the axle, the controller is able to compare the sensor values from each side and adjust braking pressure accordingly. It is believed that selecting braking pressure that gives braking for the side on ice (lowest grip) gives the best anti-lock braking performance. This is known as 'select low', and once it has happened the controller, through the solenoid

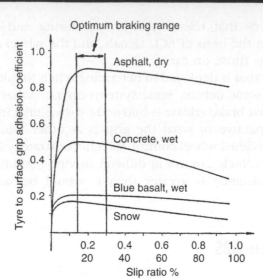

Fig. 5.6 Graph of slip ratio and coefficient of friction on different surfaces

valve, increases braking pressure on the dry tarmac side until near skid conditions are reached. In this way optimum steering control is retained together with the shortest achievable stopping distance.

In the event of any component or circuit failure, the ECU controller will automatically step down through different levels of control by switching off one or more individual wheel circuits (channels). When this happens the brakes are restored to normal manual operation and the ABS warning light is activated. In addition, the processor will register a fault code which is stored in a memory. The contents of the fault code memory can then be read out, either through the blinking light code or by connecting a diagnostic instrument to the diagnostic port. This aspect is covered in Chapter 6. Although this description is based on a Wabco design, it bears similarities to other heavy vehicle anti-lock systems. The warning light is a legal requirement for all systems as is the reversion to manual braking in the event of electronic faults.

Once again it is possible to identify the same basic elements of the system: an ECU, some sensors, some actuators (solenoid valves) and interconnecting circuits.

In all cases, whether the ABS be on a motor car or heavy vehicle, the systems are fail safe. That is to say, should the ABS system fail to operate the brakes, they will remain under the control of the driver.

As with other electronic systems we have sensors, actuators, a controller (ECU) and interconnecting circuits. All these elements are vital to the effective operation of the system. There is not much that can be done to the ECU in the garage but the other elements certainly need careful attention.

The sensors and the reluctor rotors require little maintenance, but they do require checking to ensure that the rotor is not damaged or fouled with material

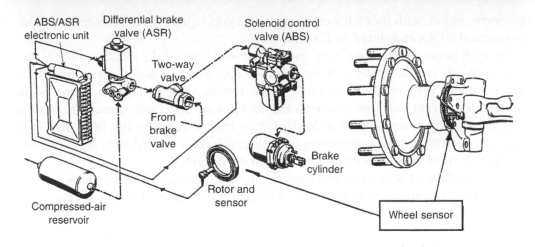

Fig. 5.7 Elements of a Wabco commercial vehicle ABS

that may affect the magnetic reluctance. This could happen if metallic particles embedded in dirt begin to fill the gaps between the teeth, or the gap between the rotor and the sensor pick-up is affected by wheel bearing adjustment: this is an area of ABS maintenance that is sometimes overlooked.

It is evident that the circuits must be in good condition otherwise the signals to and from the ECU may be impaired. Waterproof cable connectors need examination as do cable clips and cable routings to ensure that cables do not fray against moving parts. Actuators and modulators can be examined for signs of damage and general appearance.

DIAGNOSTICS

The first point of reference for diagnosis on ABS is the warning light which will be situated in a conspicuous position. When the vehicle is first started up the ABS warning light is on. The self-check capability of the ECU conducts a test of all functional circuits, solenoid valves, etc. and if all is correct the warning light goes out at a vehicle speed of about 3 mph. Should a malfunction occur during driving the warning light will come on and the ECU will initiate action to restore manual braking.

The ECU will probably contain an EEPROM (see Chapter 8), which is a non-volatile memory where fault codes are stored, or some other memory device which will permit fault codes to be read out when the system is placed in diagnostic mode.

Fault codes may be displayed at the ABS light (on-board diagnostics), or they may be read out at a diagnostic port, by a code reading tool (off-board diagnostics). In each case the diagnostic code will 'point' to an area of circuit where an error has been detected. Detailed searching is then left to the technician to perform, either with specialist equipment or a good quality multi-meter. This topic is covered in greater detail in Chapter 6.

As with other systems we have considered, the bulk of the electronics is in the ECU and this can normally only be tested by specialised means. Fortunately ECUs are normally very reliable and most of the testing that needs to be done is on components that rely on conventional technology for their operation. Sensor signals require processing in order that they can be used by the ECU, and ECU outputs usually require conversion to analogue form, coupled with amplification so that devices such as solenoid valves can be operated.

To consolidate the point about basic similarities between systems I now wish to take a look at another commonly used system, i.e. a cruise control system.

The cruise control system

Control of vehicle cruising speed is a concept that has been in use for many years. An early form of cruise control was the hand throttle. This was a lever situated at the centre of the steering column and it could be set to any position chosen by the driver so that the foot could be relieved of the effort of holding the throttle open on long journeys. The availability of electronic control of the throttle actuator means that a much more sophisticated system of control is now available. For current purposes the system described here is simplified so that essential points about structure and technology of the system can be highlighted.

The block diagram in Figure 5.8(a) shows the inputs to and output from the ECU (computer). When cruise control is required the system is energised by the main switch. The desired cruising speed is then selected and stored in the ECU memory against which road speed, as recorded by the speed sensor 5.8(b), is compared. The control strategy of the ECU will then switch the electric current supply to the actuator solenoid, 5.8(c), on or off as required so that intake manifold vacuum can be used to regulate the angle of opening of the throttle valve as shown in Figure 5.9(a) and (b).

The control rod is connected to the actuator diaphragm at one end and the throttle butterfly at the opposite end. Switching off the control current closes the vacuum valve and opens the atmospheric air valve; this permits the diaphragm control spring to push the control rod and throttle valve towards the closed position. Restoring current to the actuator solenoid opens the vacuum valve and closes the atmospheric air valve. Manifold vacuum is then applied to the diaphragm which pulls the control rod and throttle butterfly in the throttle open direction. Operation of clutch or brake pedals supplies a signal to the ECU which will cancel operation of the cruise control.

This simplified description of cruise control is given in order to emphasise certain features. (1) It contains the main elements of an electronically controlled system, i.e. sensors, actuator, ECU and interconnecting circuits. (2) The sensors use well-established electro-mechanical principles to generate electrical signals to

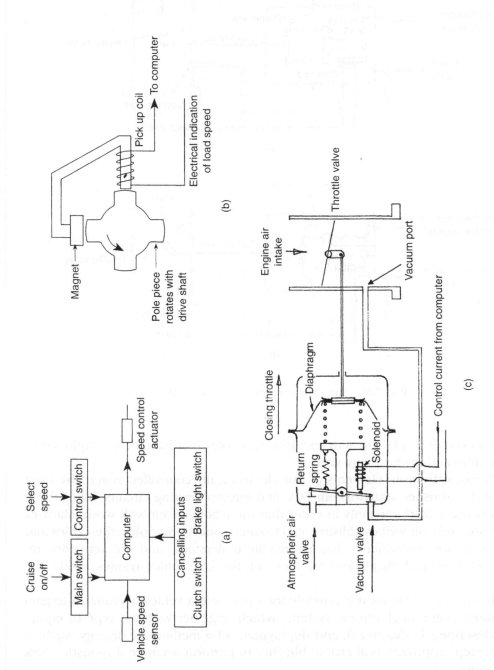

Fig. 5.8 Details of a cruise control system

Labels within figure:

(a)
Cruise on/off — Select speed
Main switch — Control switch — Speed control actuator
Computer
Vehicle speed sensor
Cancelling inputs
Clutch switch — Brake light switch

(b)
Magnet
Pick up coil — To computer
Pole piece rotates with drive shaft
Electrical indication of load speed

(c)
Engine air intake
Throttle valve
Vacuum port
Closing throttle
Diaphragm
Return spring
Solenoid
Atmospheric air valve
Vacuum valve
Control current from computer

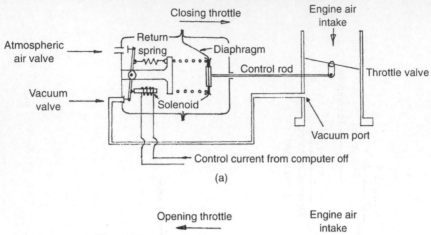

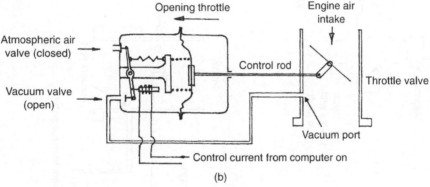

Fig. 5.9 The cruise control throttle actuator

supply to the ECU. (3) The actuator uses principles that have been employed in vehicle applications for many years.

In all cases the main elements of our electronically controlled system are seen to exist, i.e. sensors, actuators, an ECU and interconnecting circuits. All of these elements must work correctly in order that the whole system will work. Many of the sensors rely on well-established electro-mechanical principles, the cables and connectors are components that are readily understood, and the actuators are largely electro-mechanical devices. Most of the electronics resides inside the ECU.

With a bit of hard work it is possible for a competent vehicle technician to gain an understanding of electronic systems which, together with the type of equipment described in Chapter 6, and deployment of a methodical strategy, such as the 'six step' approach, will enable him/her to perform accurate diagnostic work and repairs.

In order to carry out maintenance and diagnostic checks it is necessary to be familiar with the system and this usually means that the technician will have undertaken a specialist course of training with the vehicle manufacturer. Although the operating principle of ABS wheel sensors may be similar across a range of vehicles, it remains a fact that the settings of the air gaps, and the solenoid current values, etc., may be quite different.

Accurate data that relates to a specific vehicle is an essential prerequisite for performing tests and this is often only available to approved repairers. At this juncture I wish to remind the reader that a primary function of this book is to help technicians to take full advantage of training courses that are approved by manufacturers, and this includes courses provided outside of the manufacturers' training school system.

While this may strike a slightly negative note, it should be remembered that holding firmly to the knowledge that electronically controlled vehicle systems have a common structure, namely, sensors, actuators, an ECU and interconnecting circuits, provides positive reassurance that when an organised method is used any task is manageable.

All sensor signals (inputs), all circuits, and all output devices (actuators), must be in good order otherwise the ECU will not be able to do its job. This generally requires basic electrical knowledge.

ECUs are expensive items, and changing one just to see if it cures a difficult problem can lead to disastrous consequences.

6
Diagnostics

As the use of electronically controlled systems on vehicles has increased so has the use of the term diagnostics. Diagnosis is the science, or art, of finding out what is wrong with a system and then describing the fault so that the necessary remedial work can be done. To assist those who are involved in diagnosis, a wide range of 'back up' facilities, such as tools, equipment, fault finding charts and procedures, have been developed. These 'back up', facilities and the activities involved in developing them are generally described as 'diagnostics'.

The microprocessor part of the ECU has the ability to check the circuits which are connected to it. The inputs from sensor circuits are evaluated continuously as are the output circuits to which processed data is to be sent. If an error is detected a predetermined code can be stored in a memory section which has a unique address and this encoded data can be read later, in a variety of ways, to assist in locating the cause of the defect.

The diagnostic data is accessed in two ways, one of which is on-board diagnostics and the other is off-board diagnostics. With on-board diagnostics the diagnostic data is often displayed, through a 'blinking light' system, on LEDs which are sited on the vehicle. Off-board diagnostics utilises workshop based equipment which takes a variety of forms.

In the next section an actual vehicle system is examined so that we may see how the on-board test procedure operates in practice. From this examination it should be possible to assess the skills required for diagnosis. It is important to remember that the diagnostic procedure is part of the process of gathering information which, with the aid of knowledge of vehicle systems, will enable the technician to analyse the evidence and reach conclusions about remedying defects.

We should not lose sight of the fact that we are deploying a 'six step' strategy and any system under examination should first be thoroughly examined visually to make sure that there are no obvious defects such as loose wires. It is also important, for example in the case of engine failure to start, to ensure that the ignition is working and that there is fuel in the tank. Indeed, some manufacturers of diagnostic equipment recommend that an engine 'tune' is performed first.

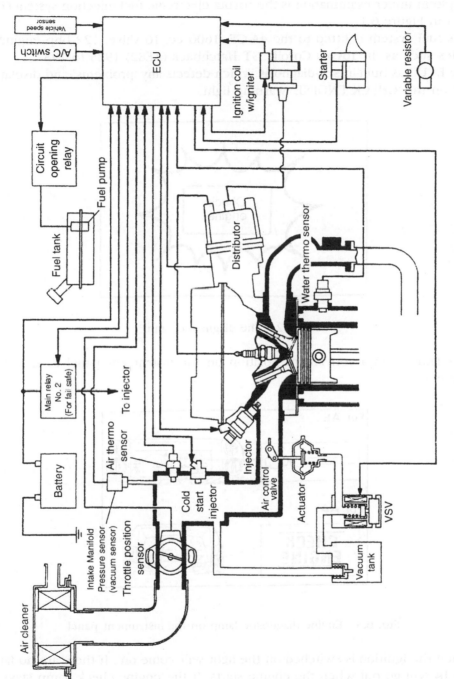

Fig. 6.1 Electronic fuel injection system (Toyota)

On-board diagnostics

The system under examination is the Toyota electronic fuel injection system (EFI) shown in Figure 6.1.

This fuel system is fitted to the 4A-GE, 1600 cc, 16 valve, 122 bhp engine in vehicles such as the Toyota Corolla GT Hatchback (AE82) 1985 to 1987.

The ECU has built-in self-diagnosis which detects any 'problems' and displays a signal on the 'CHECK ENGINE' warning light.

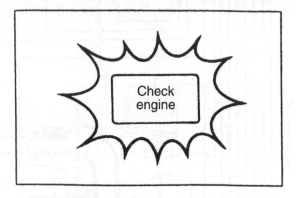

Fig. 6.2 The diagnostic lamp

This lamp is in a convenient position on the instrument panel as shown in Figure 6.3.

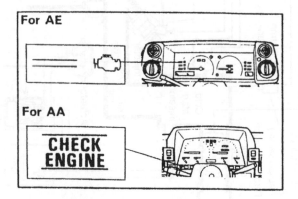

Fig. 6.3 Engine diagnostic lamp on the instrument panel

When the ignition is switched on the light will 'come on'. If there are no faults the light will go out when the engine starts. If the 'engine check' lamp stays on this is a warning that a fault is present. To find out what the fault is it is necessary to put the system into diagnostic mode. This requires some preliminary work, as follows:

1. (a) Check that the battery voltage is above 11 V.
   (b) Check that the throttle valve is fully closed (throttle position sensor IDL points closed).
   (c) Ensure that transmission is in neutral position.
   (d) Check that all accessory switches are off.
   (e) Ensure that engine is at its normal operating temperature.
2. Turn the ignition on but do not start the engine.
3. Using a service wire connect together (short) the terminals T and E_1 of the 'Check Engine' connector.
4. Read the diagnostic code as indicated by the number of flashes of the 'Check Engine' warning light.

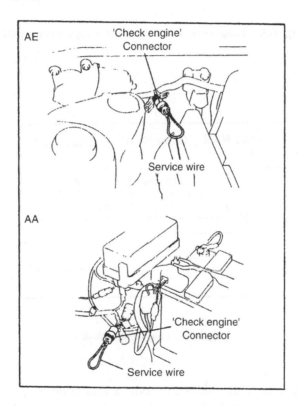

Fig. 6.4 Making the diagnostic output connection. (Note that the 'Check Engine' connector is located near the wiper motor (AE) or battery (AA), these being different vehicle models)

Diagnostic codes

Diagnostic code number 1 is a single flash every three seconds. It shows that the system is functioning correctly and it will only appear if none of the other fault codes are identified.

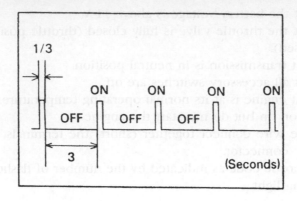

Fig. 6.5 Diagnostic code number 1 (system normal)

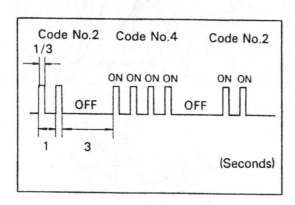

Fig. 6.6 Diagnostic codes

Figure 6.6 shows the fault codes for code 2 and code 4.

The 'Check Engine' lamp blinks a number of times equal to the fault code being displayed: there are thus two blinks close together (one second apart) for code 2, and a pause of three seconds, and then four blinks to show fault code 4. The fault code will continue to be repeated for as long as the 'Check Engine' connector terminals (T and E_1) are connected together.

In the event of a number of faults occurring simultaneously, the display will begin with the lowest number and continue to the higher numbers in sequence.

Figure 6.7(a) shows a section of the Toyota workshop manual that gives the diagnostic codes. Reading from left to right it will be seen that each code is related to a section of the system.

Code No.	Number of blinks "CHECK ENGINE"	System	Diagnosis	Trouble area	See page
1	1/3 ON →│←—ON ON ON ‖ OFF ‖ OFF ‖ OFF ‖ OFF ‖ ←3→ (Seconds)	Normal	This appears when none of the other codes (2 thru 11) are identified	–	–
2	1/3 ЛЛ ЛЛ ЛЛ Л ←1→ (Seconds)	Pressure sensor signal	Open or short circuit in pressure sensor.	1. Pressure sensor circuit 2. Pressure sensor 3. ECU	FI-31
3	ЛЛЛ___ЛЛЛ___ЛЛЛ	Ignition signal	No signal from ignitor four times in succession.	1. Ignition circuit (+B, IGt, IGf) 2. Igniter 3. ECU	FI-37
4	ЛЛЛЛ___ЛЛЛЛ___ЛЛ	Water thermo sensor signal	Open or short circuit in coolant temperature sensor signal.	1. Coolant Temp. sensor circuit 2. Coolant Temp. sensor 3. ECU	FI-35

Fig. 6.7(a) Diagnostic codes

The column marked 'See page' refers to the section of the workshop manual where further aid to diagnosis will be found. Examples of this further diagnosis may be performed with the aid of a multi-meter, utilising the knowledge of resistance values, etc. as shown in 6.7(b) and the accompanying chart. The bottom part of the diagram shows the terminals which are to be accessed in order to perform tests on circuits that the 'blink code' has shown to be defective.

Examination of the table (left-hand column) shows the test number; the next column shows the two connecting points for the voltmeter. Now take the VTA-E_2 pair of connections. With the voltmeter connected and ignition switched on the table shows that the voltage is to be read with the throttle in two positions, fully closed and fully open.

With the throttle valve fully closed, the voltage at VTA-E_2 should be 0.1 to 1.0 V. With the throttle fully open, the voltage between VTA and E_2 should be between 4 and 5 V. As a matter of interest, the reader is invited to refer back to Figures 4.24 and 4.25. It will be seen that it is, in fact, the throttle switch that is under examination.

When the diagnostic check is completed the 'Service Wire' must be removed from the 'Check Engine' connector and then the diagnostic code must be cancelled.

After the fault has been rectified the diagnostic code stored in the ECU memory must be cancelled. This is done, in the case of this Toyota model, by removing the fuse marked 'STOP', as shown in Figure 6.8. The fuse must be removed for a period of 10 seconds or more, depending on the ambient temperature, with the ignition switched off (the lower the ambient temperature the longer the period the fuse is left out).

A word of caution here. It should be noted that certain precautions need to be taken so that electronic circuits are not damaged. For example, static electricity can cause harm. It is estimated that the human body carries a static charge

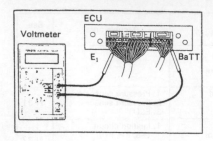

ECU
Voltmeter
E₁ BaTT

EFI SYSTEM CHECK PROCEDURE

NOTE:
1. The EFI circuit can be checked by measuring the resistance and voltage at the wiring connectors of the ECU.
2. Perform all voltage measurement with the connectors connected.
3. Verify that the battery voltage is 11 V or above when the ignition switch is at ON.

Using a voltmeter, measure the voltage at each terminal of the wiring connector.

NOTE: If there is any problem, see TROUBLESHOOTING FOR EFI ELECTRONIC CIRCUIT WITH VOLT/OHMMETER.

Voltage at ECU wiring connectors

No.	Terminals	STD voltage	Condition		See page
1	BaTT-E₁	10 – 14	—		FI-27
	+B₁ – E₁ +B		Ignition S/W ON		
2	IDL – E₂	4 – 6	Ignition S/W ON	Throttle valve open	FI-29
	VTA – E₂	0.1 – 1.0		Throttle valve fully closed	
		4 – 5		Throttle valve fully open	
	Vcc – E₂	4 – 6		—	
3	PIM – E₂	3 – 5	Ignition S/W ON		FI-31
4	FS – E₁	10 – 14	Ignition S/W ON		FI-32
	BF – E₁				
	No. 10 – E₁ No. 20	9 – 14			
5	THA – E₂	1 – 3	Ignition S/W ON	Intake air temperature 20°C (68°F)	FI-34
6	THW – E₂	0.1 – 0.5	Ignition S/W ON	Coolant temperature 80°C (176°F)	FI-35
7	STA – E₁	6 – 12	Ignition S/W ST position		FI-36
8	IGt – E₁	0.7 – 1.0	Idling		FI-37

ECU Connectors

E₀₁	No 10	STA				G(-)	G₁	IGf	T	THA	PIM	THW				Fc		BaTT	+B₁
E₀₂	No 20	IGt	E₁	Dr	V3V₁	V+	E₂₁	Ne		IDL	Vcc	VTA	E₂	VAF	FS	SPD	A/C	W	+B

Fig. 6.7(b) The ECU connections and the tests to be applied for fault tracing

equivalent to about 20 000 V. Before 'probing' into ECU connectors it is advisable to touch a good 'earth' point on the vehicle.

To summarise: the system has its own diagnostic capability. The vehicle system is prepared for outputting of the codes. The codes are read off from the warning light, as a number of blinks (four blinks for fault coded 4, etc.). Each code is separated by a pause of three seconds. When the fault codes have been recorded (on a notepad) the manual is referred to and the area of the system that is to receive further testing is shown. This 'further testing' is performed with the aid of a good quality multi-meter, the values being provided in the manual for a specific vehicle.

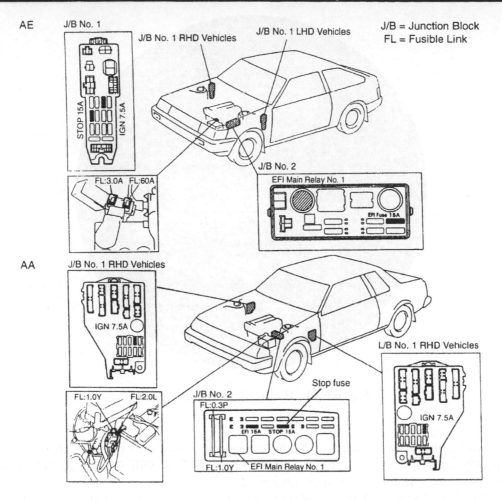

Fig. 6.8 Cancelling the diagnostic code

Diagnostics on commercial vehicles

To further assess the concept of 'on-board' diagnostics it is useful to consider the Wabco anti-lock trailer brakes diagnostics. Figure 6.9(a) shows the position of the 'blink' code LED on the ECU.

To appreciate the essential simplicity of the procedure one needs to follow the procedure through on an installation. However, a good impression may be gained from the description of Level 1 diagnostics shown in Figures 6.9(b).

Users of Wabco anti-lock systems are provided with a table of blink codes which enables them to locate the part of a circuit where a fault is indicated. For example, if the LED 'blinks' six times a fault in the circuit connected to the YE2 terminals of the ECU is indicated. This is a wheel sensor circuit.

Fig. 6.9(a)   Diagnostics LED on the ECU

The types of tests that can be performed are shown in Figure 6.10. A good quality multi-meter should be used for the tests and, as with all work on vehicles, proper safety measures must be observed.

The fault location chart also shows some tests on a modulator: this is the name given to the solenoid operated air valves. It is in fact the equivalent of an actuator in other electronic systems. However, the air brake chambers are sometimes called actuators and use of the term modulator helps to avoid confusion.

I have abbreviated the chart details in order to make a point about on-board diagnostics in general.

Please note. It must be appreciated that with this system, as with all other systems, detailed product knowledge is essential. This means that the technician must have access to the manual for the actual vehicle/trailer being worked on.

This summary of on-board diagnostics (OBD) provides an indication of the general principles that are applied. Other vehicle manufacturers use similar procedures, but the codes will probably be different, and the procedures for outputting fault codes and cancelling them quite different. However, the method of making tests at the ECU connection is fairly widely practised, and the examples shown here may be taken as indicative of circuit testing of vehicle electronic systems. The procedures are normally detailed in the manufacturer's service manuals.

Diagnostics

N.B. When diagnosing the faults via the diagnostic socket,
unless the supply is dedicated ISO 7638 only (i.e. no 24N/24S input),
the ISO 7638 power must be disconnected with the ECU powered
by the 24S/24N only.

Level 1: blinkcode - Integral LED

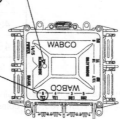

Flash code LED starts immediately a fault occurs
- single digit code flashed out
- match to figures marked on ECU housing.
         example:LED flashes 6x (6 times)
             Faulty component is YE2
             Sensor and/or cable from
             socket YE2 is faulty

Level 2 - blinkcode plug / hand held tester   (full instructions supplied with each unit)

blinkcode plug
446 300 334 0

Normal mode: Press key on blinkcode plug once. Output same as level 1 diagnostic
Expert mode : Press key on blinkcode plug twice.System recognition and stored faults indicated.

example:

               short circuit

         activate   4S/2M   sensor BU1   occurred twice

Additional functions: System set - allows installed system to be altered i.e. 2S/2M system reset to 4S/2M
                        System test - check correct pneumatic and electrical allocation after initial installation
                        or repairs

LED digital output - no need to count flashes

Displays: system fitted   e.g. *UCS* = VCS
              error codes   e.g. *4 71* = sensor D (YE1) air gap too big, occurred once
              system configuration e.g. *4 - 2* = 4S/2M system

hand held tester
446 300 400 0

Special functions: System initialisation - allows installed system to be altered
                        i.e. 2S/2M system reset to 4S/2M
                        Function test - check correct sensor / modulator matching and operation
                        Mileage counter
                        Speed dependent switch output
                        Service signal

Level 3 - Diagnostic controller + smart card

Diagnostic controller
446 300 331 0

Enables a full and comprehensive diagnosis to be carried out,
controller also includes a multimeter.
On-line help screens guide the use through the test steps.
With an appropriate printer test report can be printed out.

Fig. 6.9(b)   The Wabco 'blink' code systems

As will be seen in the next section, which deals with off-board diagnostics,
testing the circuit at the ECU may only be the beginning of the process because
of the circuit elements, such as connectors, relays, and so on.

Fault Location

The following should be noted: (refer to wiring above)

system	modulator			sensor			
	RD (A)	YE (B)	BU (C)	BU1 (C)	BU2 (E)	YE1 (D)	YE2 (F)
2S/2M		✓	✓	✓		✓	
4S/2M		✓	✓	✓	✓	✓	✓
4S/3M	✓	✓	✓	✓	*	✓	*

* BU2 and YE2 sensed wheels controlled by RD modulator

Important note: In the following examples, before carrying out any tests, switch off the power and always ensure that the Sensor/Modulator connections, if removed, are replaced back in exactly the same locations.

Sensors

Large air gap:
Example: Wheel E (BU2)

remove sensor connector BU2	jack up wheel and turn by hand	measure sensor output min 0.1v AC	reset sensor gap

Speed drop out / Intermittent speed / Speed jump:
Example: wheel C (BU1)

remove sensor connector BU1	jack up wheel and turn by hand	measure sensor output max output < 2x min output	**check for:** excessive pole wheel run-out (max 0.2mm TIR) / damaged teeth / loose wheel bearings / security of sensor mounting

Broken wire/short circuit
Example: Sensor D (YE1)

remove sensor connector YE1	check sensor resistance should be 1.0/1.3 Kohm	if not OK disconnect cable from sensor	check resistance directly to identify location of fault	if OK

reconnect sensor to sensor cable	check for breakdown of the insulation between sensor wiring and chassis - should be > 100,000 ohms if fault - use previous method to isolate faulty component

Modulators

Example: Broken wire Mod C (BU)

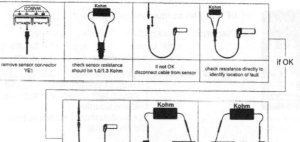

remove 'Y' connector top RHS of ECU	check resistance: outlet solenoid & inlet solenoid	check resistance between pins 1 & 2	if open or short circuit - check resistance at modulator to determine location of fault

Power supply wiring:
These faults are usually caused by blown fuses, damaged wiring or incorrect connections at the headboard sizes or chassis junction boxes - check by normal continuity/short circuit checks. In many instances a test bulb (24v / 5w) is better than a multimeter especially where high resistance joints are concerned.

Fig. 6.10 Fault location and circuit tests

Off-board diagnostics

This term applies to the type of diagnosis that is performed in the workshop and requires the use of dedicated test equipment. The range of such equipment is vast, from personal computers with peripherals to relatively simple hand-held testers. It is the hand-held type of tester that I have chosen to describe because it is easy to use and not too expensive to buy, and it gives a good insight into the general principles that lie behind off-board diagnostic practices.

I am indebted to former colleagues of Brighton College of Technology and Eastbourne College of Arts and Technology who kindly discussed with me the off-board diagnostic equipment they use for teaching purposes, and also for permitting me to take photographs of them at work.

Fig. 6.11   The off-board diagnostic kit

Figure 6.11 shows the principal parts of the diagnostic kit. There is a hand-held tester, a lead to connect the tester to the vehicle's diagnostic connector, a 'smart card' that matches the tester to the vehicle system under test, a printer to provide a permanent record of the test results, and leads for making connections to the battery and from tester to printer. This is accompanied by the all-important instruction manual (Fig. 6.12), although it needs to be stated that once the test program has started, the display screen on the tester provides a step-by-step menu to guide the operator through the test sequence. The 'smart card' is the equivalent of computer software and it enables the tester to utilise the ECU processor power to interrogate circuits. The test instrument is thus able to test all circuits that are served by the ECU; the test connection plug is also known as the serial port because test information is fed out serially (one bit after the other, e.g. 10110011).

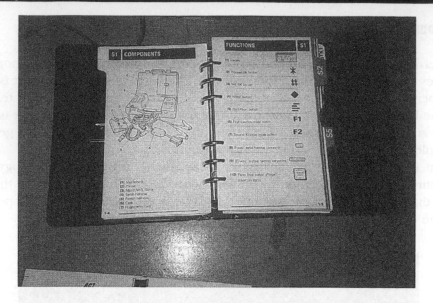

Fig. 6.12 The instruction manual

A considerable advantage of the serial port is that it permits testing without the need to disconnect wiring.

Mention has been made earlier about the need for a location chart which shows the position of the various elements of the electronic system. Unless the operator is familiar with the vehicle, it will be necessary to refer to this chart in order to locate the diagnostic connector. Figure 6.13 shows the Rover 200 series connection point. This was the vehicle on which my tests were performed.

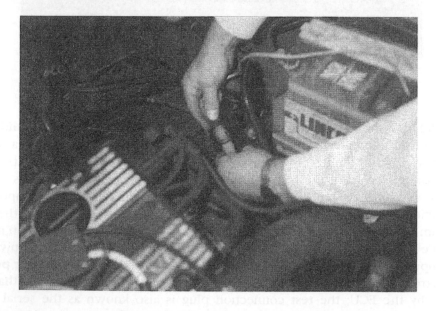

Fig. 6.13 The diagnostic connector

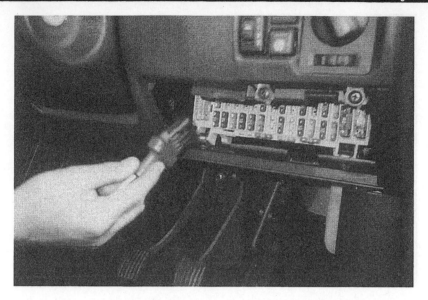

Fig. 6.14 The equivalent connection on the Vauxhall Tigra

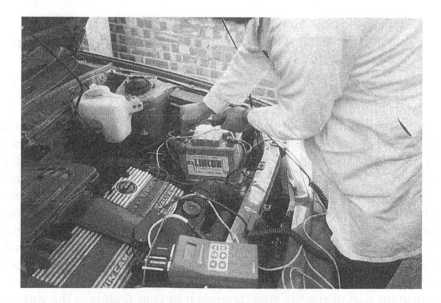

Fig. 6.15 Connecting to the power supply

Figure 6.14 shows the position of the diagnostic connector for my friend's Vauxhall Tigra. The one in Figure 6.13 is in the engine compartment and the Tigra one is inside the car, under the dashboard. This illustrates the importance of having information that relates accurately to the vehicle being tested.

The source of power for the tester is the vehicle battery, and Figure 6.15 shows the leads being connected. The tester is placed in the position shown for the sole

Fig. 6.16   Inserting the 'smart' card to adapt the test instrument to the specific vehicle application

purpose of taking the photograph. For test purposes, it is held in the hand and it should be noted that care must be taken to place the instrument in a safe position when not being held in the hand.

The next step is to connect the diagnostic lead to the vehicle's diagnostic connector, and Figure 6.13 shows this being done.

Before starting the test, the diagnostic lead that relates to the specific vehicle model will have been selected, as will the 'smart' card that customises the test instrument to the vehicle. Figure 6.16 shows the 'smart' card being inserted.

It will be understood that different makes of vehicles require different types of diagnostic leads and 'smart' cards. This is necessary because the diagnostic connections vary from vehicle to vehicle, as does the test program. Figure 6.17 shows the diagnostic lead and 'smart' card that customises the test instrument to a Ford vehicle. A reasonable range of diagnostic leads and 'smart' cards is available to make this type of equipment suitable for use on a number of vehicle makes. It will be appreciated that this is an important consideration for the independent garage which is not linked to a particular vehicle manufacturer.

An important part of the 'six step' approach to fault finding is the gathering of evidence. Figure 6.18 shows the printer being connected, for it is from the printout that a permanent record of the test results will be obtained – as shown in Figure 6.19(a).

When all leads have been correctly connected and steps taken to ensure that leads are clear of drive belts, hot engine parts, etc., the test may commence. The manual (Fig. 6.12) gives a description of the instrument controls, and when all preparations are made the test instrument screen displays a message which guides the operator through the test sequence.

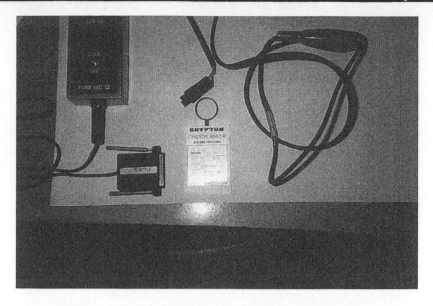

Fig. 6.17 The diagnostic lead and 'smart' card for a Ford

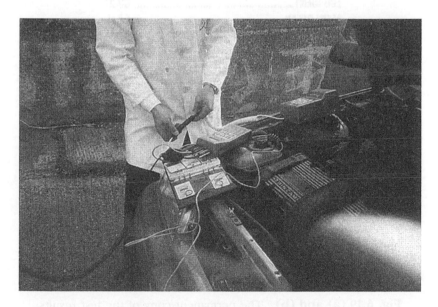

Fig. 6.18 Connecting the printer

The test procedure may require the operator to operate certain of the vehicle controls. Figure 6.20 shows the accelerator being depressed. Here it will be seen that a certain amount of movement around the vehicle is required during a test sequence. It is thus important that care is exercised to ensure that leads do not

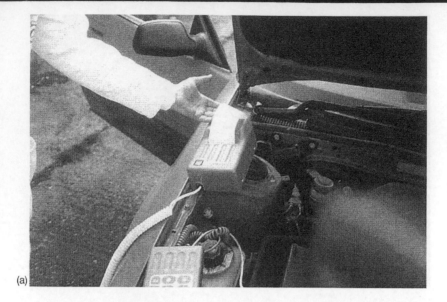

(a)

(b)

Test Results

BATTERY VOLTAGE...GOOD
SERIAL COMMS. ....GOOD
ECU IDENT ....GOOD
MAIN RELAY ....FAIL
COOLANT CIRCUIT ....GOOD
INJECTORS ....GOOD
FUEL PUMP CIRCUIT ....GOOD
THROTTLE SWITCH CIRCUIT ....GOOD
THROTTLE POT CIRCUIT ....GOOD
OXYGEN SENSOR HEATER RELAY ....GOOD
PURGE VALVE SOLENOID ....GOOD
VACUUM SENSOR/PIPES ....GOOD
STEPPER MOTOR CIRCUIT..............................GOOD
CRANK SENSOR CIRCUIT.................................GOOD
IGNITION LT CIRCUIT ....GOOD
HT/FUEL ....GOOD

PRINT NUMBER 7

Fig. 6.19 (a) and (b) The permanent copy of the test results

become entangled and that the vehicle is positioned so that freedom of movement around it is ensured.

Previous examination of other forms of diagnostic equipment shows that a small trolley (Figure 6.21), on which to place instruments, manuals and tools, is a

Fig. 6.20 Operating the vehicle controls during the tests

useful aid to electronic fault diagnosis. For, as you will understand, it is not always easy to find a position on the vehicle where test equipment can be safely placed.

On completion of the test the printout is produced, as shown in Figure 6.19, and analysis of the results can proceed. In this particular test, the printout (Figure 6.19(b)) recorded that the 'Main relay' was faulty. This highlights a feature of diagnostics that has been a constant theme in this book, which is that the diagnostic equipment directs the technician to the 'area' of the fault, and once that has been done he/she should resort to basic electrical/mechanical knowledge and skills. These basic skills are a necessary condition for a satisfactory understanding of vehicle work.

In the case of the main relay, it must first be located. This requires the use of a location chart (Figure 6.23) which shows the position of the device on the vehicle. It is then necessary to gain access to it and to perform a visual inspection. It may be a loose connection. It is tempting to just fit a new relay, but this should never be done until the device has been proven defective, and if we stick to our 'six step' approach we shall determine the cause of failure, repair this cause so that it does not produce a repeat failure, and then carry out a test.

For reasons of cost and physical limitations, the fault tracing capacity of the ECU and the fault code system is generally limited to identifying the area of the system (circuit) in which a fault has been detected.

Let us assume that the diagnostic tester gives a message

'FUEL PUMP CIRCUIT – – – – FAIL'

This does not mean that the fuel pump has failed. It means what it says: the fuel pump circuit is not working correctly. To find out what is wrong, it is necessary

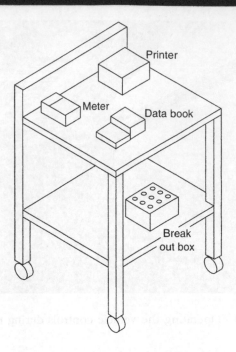

Fig. 6.21 Trolley for diagnostic kit

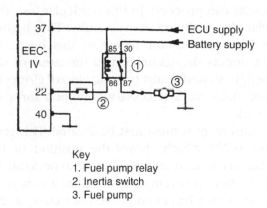

Key
1. Fuel pump relay
2. Inertia switch
3. Fuel pump

Fig. 6.22 Fuel pump circuit (Ford Escort/Orion)

to know what the circuit consists of. This information is provided in the instrument instruction manual. In this particular case the circuit diagram is for the Ford Escort/Orion models.

It is now possible to see that although the diagnostic tester has taken us part of the way there is still quite a bit of testing to do. In addition to the relay, the pump motor and the inertia switch, there is a fuse. There are also several connectors and the cables themselves. Any of these items may be at fault, but some are more likely causes than others.

In keeping with our systematic 'six step' approach, we should not jump to conclusions. However, before we consider the testing of the circuit we will need to know where the circuit elements are located on the vehicle and, as you are probably aware, the circuit diagram only tells you what is in the circuit: it does not tell you where the various items are located on the vehicle. This is where the location chart (Figure 6.23) comes in useful.

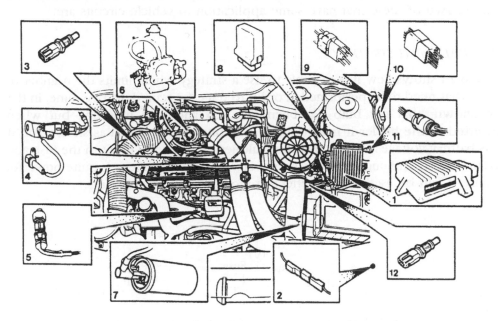

1. ERIC ECU
2. Ambient air temperature sensor (located behind horns)
3. Coolant temperature sensor
4. Crankshaft sensor
5. Knock sensor
6. Carburetter
7. Ignition coil
8. Engine MFU
9. 4-way connector—engine/main
10. 13-way connector—engine/main
11. Serial Diagnostic Link connector
12. Inlet air sensor

Fig. 6.23 A typical location chart

Although this chart refers to a different vehicle it does give an idea of the way that the diagnostic system, in its entirety, works – which is the purpose of the exercise. In this instance, it gives the name of the component, a picture of the unit and a clear indication of its position on the vehicle.

In this chapter we have given fair consideration to the principles involved in fault diagnosis. In the cases of 'blinking LED' codes as displayed by the hand-held instrument and similar equipment, the general area of the fault is detected. From that point it is usually necessary to resort to the use of a multi-meter. A digital type meter is best for this purpose because its electrical characteristics match those of the circuits being tested.

It is obvious that the multi-meter is only of use if the operator knows what to do with it, and this is where knowledge of circuits, sensors and actuators plays a part.

If we now refer back to our fuel pump circuit we shall see that it is composed of fairly standard electrical devices and each of these may be tested with the aid of a good quality multi-meter. Of course, if it is a fuse that has blown, or an inertia switch that has tripped, these will have been checked at an early stage to save unnecessary work. In this connection it is useful to note that there are techniques for searching circuits which, when used correctly, can lead more quickly to a solution. Two of these that have some application to vehicle circuits are:

1. THE HALF SPLIT STRATEGY

Figure 6.24 shows how it is possible to use a strategy which limits the number of checks that need to be performed in order to trace a circuit defect. Assume, in the case shown, that the bulb has been checked and it is in order, but when reinserted in the holder it fails to light. If a voltmeter is placed as shown and it reads battery voltage it shows that the fault lies between the input to the fuse and the lamp earth. Although this may seem very simple the strategy is quite powerful because it can be used to good effect in more complicated circuits.

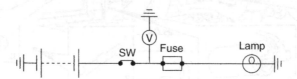

Fig. 6.24   The principle of the half split method

2. HEURISTICS

Another name for heuristics is 'rule of thumb'. As an example, assume that there is a vehicle known as the Xmobile. This vehicle is notorious for starting difficulties in damp, misty conditions. It becomes known that drying out HT insulations is often beneficial for producing a start. In the case of a 'call out' for failure to start which involves one of these vehicles a sensible use of the 'rule of thumb' would be to take account of the weather conditions and to 'dry out' the HT insulation before attempting anything more elaborate.

The manual contains circuit diagrams and test suggestions for each type of system (smart card and diagnostic connector) that the user has selected for their tester. These circuit tests can be performed with the aid of a multi-meter and considerable skill on the part of the technician.

As shown in the introduction, the use of electronics continues to grow and it is reasonable to expect diagnostic systems to grow proportionately. This does seem to be the case, and manufacturers and suppliers of test equipment often have a telephone 'help line' through which users may gain assistance. Further developments have led to manuals and fault finding guidance being available on CD ROM and, as systems develop and are networked on the vehicle, it becomes more important to have a standardised tester interface so that all systems on the vehicle may be accessed through the interface and a serial data line. More will be said about this in Chapter 7.

The complexity of the circuit is related to the function that it has to perform, but it is assumed that it will contain a sensor, some cables and connectors, and an actuator, and probably some relays. It is important to have an understanding of these devices in order to be able to test them. If we refer back to our fuel pump circuit we shall see that it contains a range of such units and this is a convenient point at which to consider them in detail.

From this fuel pump circuit it is possible to choose a selection of circuit elements which are commonly used, such as relays. The relay is an electro-magnetic switch in which a relatively small current can be used to switch on a much larger current. This means that the switching device and the cable carrying current to the relay carry a much smaller current than is required by the device that the relay contacts operate. The circuit carrying the larger current can be made smaller (shorter) as a result.

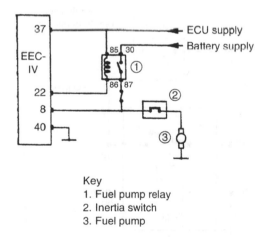

Key
1. Fuel pump relay
2. Inertia switch
3. Fuel pump

Fig. 6.25 Fuel pump circuit components (Mondeo)

Relays are used extensively in electronically controlled systems. The principle of the relay is that a coil provides a magnetic field when current flows through it. Adjacent to the coil is a movable lever (armature) which is attracted towards the coil by the magnetic field. An electric switch contact is attached to the armature and this contact is mated with another, to complete the circuit, when the coil is energised. When current is removed from the coil a spring will cause the switch contacts to separate and open the circuit. This action refers to a switch that is normally open. Some relay switches are normally closed and in such cases the current in the relay coil will cause the switch contacts to separate.

Figure 6.26(a) shows the general principle of the relay; it is fundamentally quite simple and can be tested by simple means. For example, the coil *can be* checked

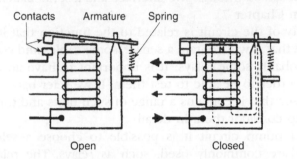

Fig. 6.26(a)   The principle of the relay

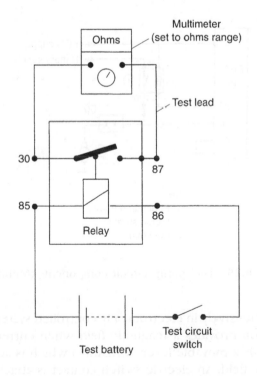

Fig. 6.26(b)   A relay test

for continuity (i.e. is the coil circuit complete?). If the coil *is* complete, the relay can be energised by applying 12 volts to the coil, provided it is a 12 volt system. Energising the coil should operate the contacts and there should then be continuity between the terminals which relate to the switch contacts.

Great care should be taken when performing such tests because a 12 volt battery contains a great deal of electrical energy and an accidental short circuit can have serious effects such as fire and personal injury.

Figure 6.26(b) shows the type of circuit arrangement that could be used for testing a simple relay. The use of a switch means that all the test connections can be made securely before the switch is closed to energise the relay coil. When the switch is closed, there should be virtually zero resistance between terminals 30 and 87.

Figures 6.27(a) and (b) show the circuits of two commonly used types of relay. These diagrams give an insight into the types of tests that can be applied in order to verify the condition of a relay. For example, type b, the switching relay, will require continuity through the electro-magnet and through the common contact to one of the two output terminals when the coil is energised, and to the other output terminal when the coil is not energised.

These diagrams also show that relays of different types are to be found on vehicles and, because of their differing functions, they are not interchangeable. The function of the resistor which is connected in parallel with the relay coil (Fig.

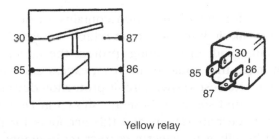

Yellow relay

Fig. 6.27(a)

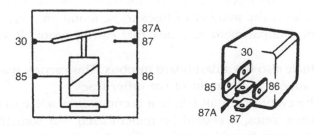

Fig. 6.27(b)

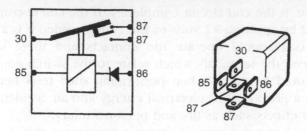

Fig. 6.27(c)

6.27(b)) is to protect the circuit against back emf (Lenz's law) that arises when the relay is switched off.

Figure 6.27(c) shows a relay fitted with a circuit protection diode. It is evident that relay energising current will only flow when the terminals 85 and 86 are connected to the correct polarity, which means – as ever – that accurate knowledge of the system and its components is essential.

Relays are often mounted together in one position on the vehicle, as shown in Figure 6.28.

Circuit protection

The sub-circuit chosen for this examination contains a fuse. The purpose of the fuse is to provide a 'weak' link in the circuit which will fail (blow) if the current exceeds a certain value and, in so doing, protect the circuit elements and the vehicle from the damage that could result from excess current.

The fuse is probably the best known circuit protection device. There are several different types of fuses and some of these are shown in Figure 6.29.

Fuses have different current ratings and this accounts for the range of types available. Care must be taken to select a correct replacement; larger rated fuses must never be used in an attempt to 'get round' a problem. Many modern vehicles are equipped with a 'fusible link' which is fitted in the main battery lead as an added safety precaution.

It is common practice to place fuses together in a reasonably accessible place on the vehicle. Another feature of the increased use of electrical/electronic circuits on vehicles is the number of fuses to be found on a vehicle. Figure 6.30 shows an engine compartment fuse box that carries fusible links in addition to 'normal' fuses.

The same vehicle also has a dashboard fusebox. This carries 'spare' fuses which, in Figure 6.31, appear to the right of the other fuses.

Whereas in the event of circuit failure it is common practice to check fuses and replace any 'blown' ones, it should be remembered that something caused the fuse to blow. Recurrent fuse 'blowing' requires that circuits should be checked to ascertain the reason for the excess current that is causing the failure.

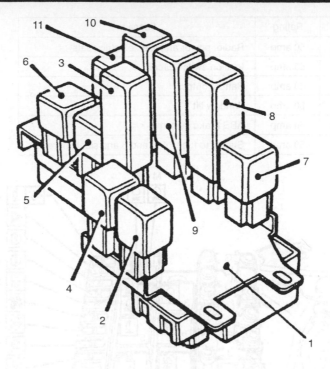

1 Relay panel
2 Front window lift relay
3 Interior light delay unit
4 Fuel pump relay
5 Rear foglamp relay
6 Side lights relay
7 Rear window relay
8 Headlamp delay timer
9 Headlamp changeover unit
10 Heated rear window timer
11 Wiper motor relay

Fig. 6.28 A relay panel (Rover 800)

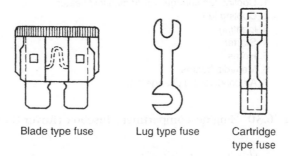

Blade type fuse   Lug type fuse   Cartridge type fuse

Fig. 6.29 Fuse types

Although it is not shown in our sub-circuit of the fuel system, some circuits are protected by thermal circuit breakers, and this is a convenient point at which to introduce the circuit breaker.

No.	Rating	Function
G	50 amp	Radio, power amplifier, electric seats
H	50 amp	Ignition switch circuit
I	80 amp	Battery output
J	50 amp	Window lift
K	50 amp	ABS brake system
L	50 amp	Supply to fuses 4, 5 and 6 and sidelight relay

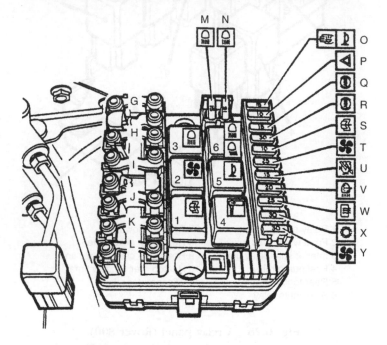

Relays
1. Cooling fan changeover or manifold heater
2. Cooling fan
3. Lighting
4. Starter
5. Horns
6. Main/dip beams
7. Air conditioning changeover

Fig. 6.30   Engine compartment fusebox (Rover 800)

The thermal circuit breaker relies for its operation on the principle of the bi-metallic strip. In Figure 6.32 the bi-metal strip carries the current between the terminals of the circuit breaker. Excess current, i.e. above that for which the circuit is designed, will cause the temperature of the bi-metal strip to rise to a level where the strip will curve and allow the contacts to separate. This will open the circuit and current will cease to flow. When the temperature of the bi-metal strip falls the circuit will be re-made. This action leads to intermittent functioning

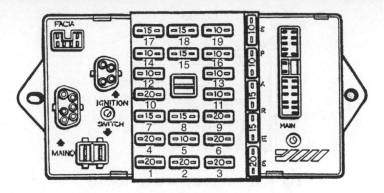

Fuse functions

Fuse No.	Rating	Wire colour	Function
1	20 amp	N/O	Sunroof, driver's seat heater
2	20 amp	S/U	RH door window lift
3	20 amp	S/R	LH front door window lift
4	20 amp	P/N	Front and rear cigar lighters Footwell lamps
5	10 amp	P	Interior lights, boot light, interior light delay unit, map light, door open lights, trip computer memory, radio station memory, clock memory, headlight delay unit
6	20 amp	P/O	Burglar alarm ECU (optional) Central locking ECU
7	15 amp	R/O	RH number plate lamp RH side, tail and marker lights Trailer plug
8	10 amp	R/B	Cigar lighter illumination, LH licence plate lamp, sidelight warning light, glovebox light, trailer plug, dashboard illumination, LH side tail lights
9	20 amp	S/O	LH rear window lift
10	20 amp	S/G	RH rear window lift

Fig. 6.31 Dashboard fusebox (Rover 800)

of the circuit, which will continue until the fault is rectified. An application may be of a 7.5 ampere circuit breaker to protect a door lock circuit. The advantage of the circuit breaker over a fuse is that the circuit breaker can be re-used.

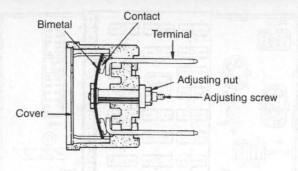

Fig. 6.32   A circuit breaker (Toyota)

CONNECTORS

Greater use of electrical/electronic circuits has led to an increase in the number of electrical connections. These connections may consist of multiple pins, as in the case of the main wiring harness at the ECU, or a single connection, as in the case of an ignition coil.

Comprehensive coverage of connectors is beyond the scope of this book, but it is helpful to consider some of the basic principles because connectors are

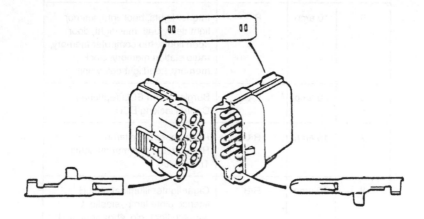

Fig. 6.33   Multiple pin connector

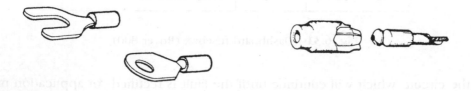

Fig. 6.34   Single cable connector

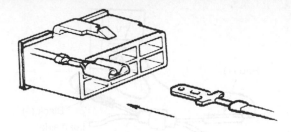

Fig. 6.35   Checking the spring force of a connector (Toyota)

thought to be a major source of problems if they are incorrectly fitted or poorly maintained.

Connectors of the type shown are commonly used in vehicle circuits. The quality of the electrical connection that is made is largely dependent on the spring force incorporated into the design of the connector. Should the spring weaken the electrical connection will be impaired and arcing may occur. This can lead to complete failure of the device which is being supplied via the connector.

As shown in Figure 6.35, the tightness of the spring force in the connector may be assessed by carefully pushing the male element into the female one. Should there be little resistance to this action it is recommended that the female part of the connector be replaced.

If it is possible to gain access to the connector as seen in Figure 6.36(a), a simple resistance test will show the condition of the connector. This test can only be performed when there is no current in the circuit. Figure 6.36(b) shows a voltage drop test being conducted. In this case the circuit would be switched on and the voltage drop measured should be virtually zero.

Because connector contacts may be affected by moisture, waterproof connectors have been developed to reduce the possibility of such damage. Care needs to be exercised when dealing with this type of connector so as not to damage the water seal. Some manufacturers recommend the use of a special grease to reduce the risk of corrosion at connector contacts. Figure 6.37 shows a waterproof connector.

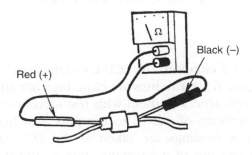

Red (+)

Black (−)

Fig. 6.36(a)   Ohm-meter test on connector (Toyota)

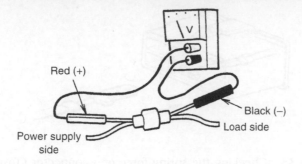

Fig. 6.36(b)   Voltage test on a connector

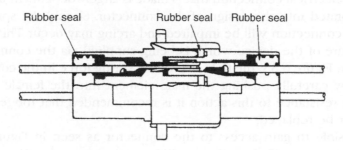

Fig. 6.37   Waterproof connector (Toyota)

THE WIGGLE TEST

Connectors are often thought to be the cause of intermittent faults on electronic systems. As the name implies, the wiggle test requires that the two sides of a connector are moved relative to each other. If this test is applied with reasonable force any defect that only occurs under movement should be picked up. The ECU will detect a fault and, by use of the diagnostics, one should be able to detect intermittent faults arising from loose connector contacts.

THE 'BREAKOUT' BOX

The cables that connect a circuit to the ECU do so through a multiple-pin plug of the type shown in Figure 6.39. The pins are close together and quite small, which makes it difficult to gain access to them with test meter probes.

To overcome the problem of access to the pins, and because it is generally recommended that test readings are taken at the ECU end of cables, some manufacturers prefer the use of a 'breakout' box of the type shown in Figure 6.40.

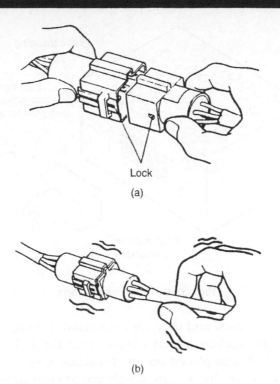

Lock

(a)

(b)

Fig. 6.38(a) and (b) The 'wiggle' test

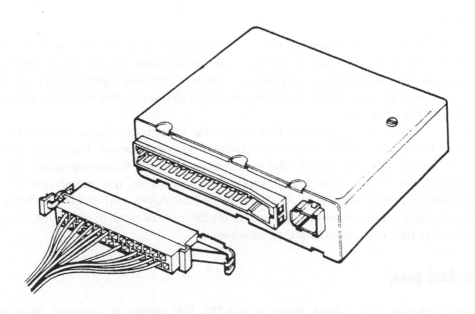

Fig. 6.39 A multi-pin ECU connector

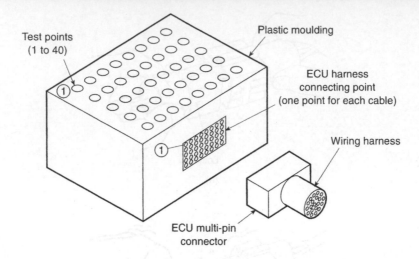

Fig. 6.40 A 'breakout' box

The test points are robust and they are connected, through the 'breakout' box, to the connector pin sockets which couple with the ECU socket, when it is removed from the ECU and placed on the 'breakout' box.

The large test points on the box are numbered to correspond to the same pins on the ECU connector, and these are the numbers that are shown on the circuit diagram. It is common practice for manufacturers to give test data in terms of readings taken at the ECU connections, so the better the access to these pins the more reliable are the readings likely to be.

SAFETY

One's personal safety must always be safeguarded, as must the safety of other workers, so it is essential to observe all precautions. At the same time it is important not to damage components. Many electronic components are very sensitive and it is advisable that only good quality (probably digital) meters are used for test purposes.

There is also the possibility that the static electricity charge carried by the human body (20 000 V) may cause damage to components. I recently fitted a new component to my PC and the instructions gave the following advice: 'Before handling the PCB, touch a grounded object, such as the metal case of the computer's power supply. Hold the PCB by the edges and avoid touching the components.' I am assured by a major manufacturer of vehicle electronic components that this is good practice to observe.

The fuel pump

Before leaving this section, there is still the fuel pump to consider. Bearing in mind that procedures will vary from one type of vehicle to another I give the

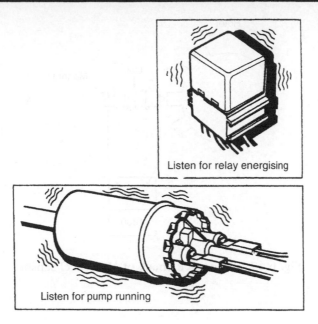

Fig. 6.41   The fuel pump 'hums' when it is running

following excerpt from a Rover manual. 'Listen for pump running. If the fuel pump is not running, check that the pump relay is operating. If OK, check the wiring from the relay through the ballast resistor to the pump and the wiring from the pump to earth. If OK, renew the pump and retest.'

Oscilloscope testing

Machines such as the Crypton CMT2000 IT workstation have been a familiar sight in workshops for many years and their use for condition analysis of vehicle systems is an asset to vehicle technicians. The use of such equipment for engine condition, ignition circuit and charging system tests is well understood. Oscilloscopes are also useful for performing more searching tests on inductive type circuits, such as ABS wheel sensor windings.

Although a resistance check of a coil winding will provide a good indication of the condition of a coil operated (inductive) type device, it does not always provide a conclusive test result because the resistance may be affected by the operation of the sensor. A dynamic test as can be performed with the oscilloscope, while the vehicle is running, can provide more conclusive information.

For example, consider the ohm-meter test on a wheel sensor winding. The wheel sensor shown in Figure 6.42, diagrammatically, has a resistance which must lie between the limits of 1600 ohms and 2100 ohms, as measured at the ECU input.

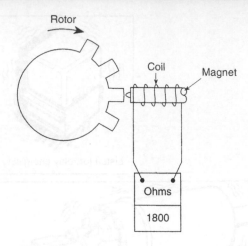

Fig. 6.42 Resistance check of ABS wheel sensor

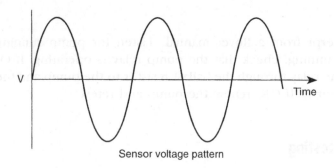

Fig. 6.43 Voltage waveform for ABS wheel sensor

This check will verify the condition of the sensor coil but it will not tell us anything about the output from the sensor when the vehicle is in motion. For example, should the air gap between the sensor pick up and the rotor be incorrect the output of the sensor will be affected. However, an oscilloscope connected to the same two terminals could be used to obtain a picture of the output if, for example, the vehicle were to be operated on a rolling road dynamometer. With an oscilloscope suitably connected to a wheel sensor the type of voltage waveform shown in Figure 6.43 would show that the sensor was functioning correctly, under operating conditions.

It is now possible to obtain test instruments such as the Crypton CPT50 portable scope that should make it easier to perform such tests.

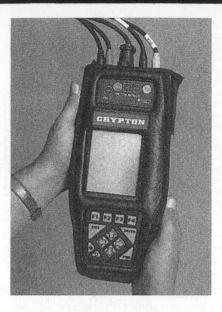

Fig. 6.44 The Crypton CPT50 Portable 4 Channel Tech Scope

Fault tracing aids

CIRCUIT (WIRING) DIAGRAMS

The wiring diagram is an essential aid to fault tracing. To help the user understand a wiring diagram certain codes are used. Two important codes that are used for this purpose are:

1. The colour code.
2. The code of symbols used to represent devices as shown in Chapter 3, Figure 3.5(b) and (c).

1. The colour code.
   Commonly used colour codes are:

N = brown	Y = yellow
P = purple	K = pink
W = white	R = red
O = orange	LG = light green
U = blue	B = black
G = green	S = slate

   To assist in tracing cables they are often provided with a second colour tracer stripe. The wiring diagram shows this by means of letters, e.g. a cable on the wiring diagram with GB printed on it is a green cable with a black tracer stripe. The first letter is the predominant colour and the second is the tracer stripe. Note,

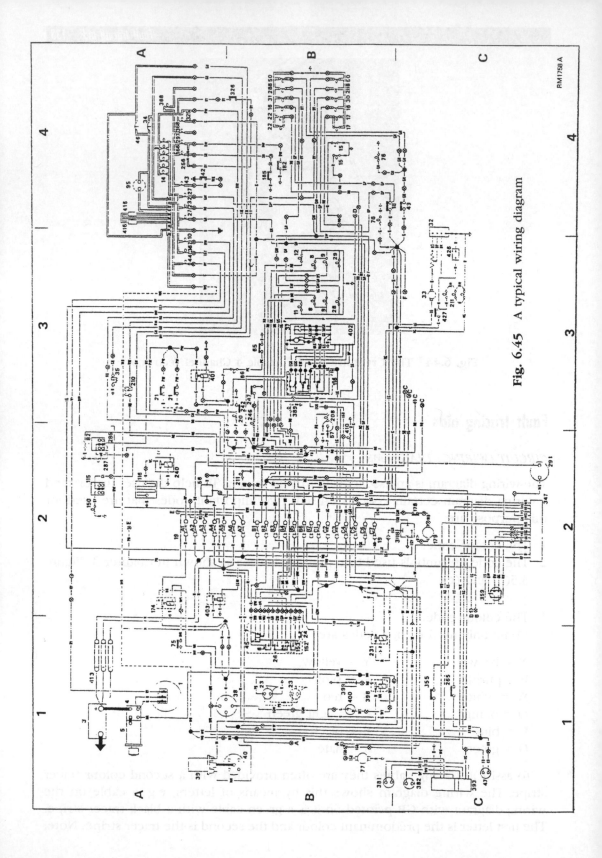

Fig. 6.45 A typical wiring diagram

RM1758 A

however, that this is not a universal colour code and, as with many other factors, it is always wise to have accurate information to hand that relates to the product being worked on.

The predominant (main) colours frequently relate to particular circuits as follows:

- Brown (N) = Main battery feeds
- White (W) = Essential ignition circuits (not fused)
- Light green and also green (LG) (G) = Auxiliary ignition circuits (fused)
- Blue (U) = Headlamp circuits
- Red (R) = Side and tail lamp circuits
- Black (B) = Earth connections
- Purple (P) = Auxiliary, non-ignition circuits, probably fused.

Figure 6.45 shows a full wiring diagram for a vehicle. To make it more intelligible Rover uses a grid system. Numbers 1 to 4 across the page, and letters A, B, C, at the sides. This means that an area of diagram can be located. For example, the vehicle battery is in the grid area 1A.

As a further aid for the user, wiring diagrams are often broken down into circuits that relate to a specific system. An example is shown in Figure 6.46. Should the diagnostics report an injector circuit fault, reference to the code shows that the injectors are at 20, 21 and they are connected to the main relay by a YB (yellow with black tracer) cable. As shown in the description of the ohmmeter test for the resistance of the injector coil and the separate resistor, this diagram directs the technician to the precise part of the circuit where the tests are conducted.

Other fault tracing aids

Fault tracing information may be provided in many different forms. Two of these are: (1) the decision flow chart (algorithm); and (2) the fault finding chart (trouble shooting chart). Examples of these are shown in Figures 6.47 and 6.48.

It must be understood that the aids shown here are related to specific vehicle models. Other vehicles will have different symptom–fault linkages, and test values will be different. However, they serve to highlight the fact that such aids exist and that they can play a constructive part in the process of searching for faults in a sub-system after diagnostic tools have directed the technician to the sub-system.

To further assist the user of a wiring diagram, components, motors, injectors, relays, etc. are usually identified by a number. This number is listed on the diagram, or close to it, together with a description of the component. This is important because, rightly or wrongly, there is some variation to be found in the symbols that are used to represent a device as, for example, may be seen in Chapter 3, Figure 3.5(b).

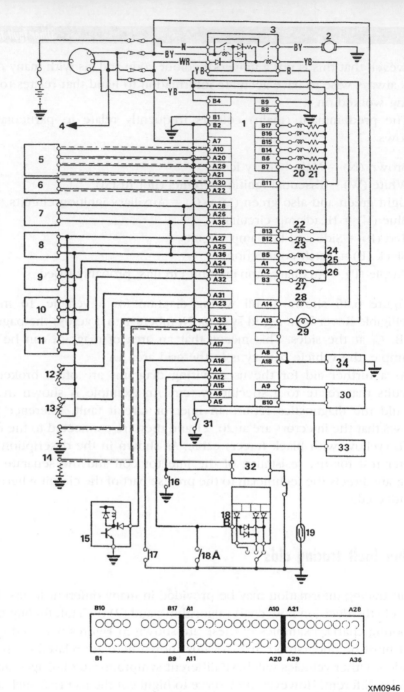

XM0946

1. ECU	13. Air Intake Temperature Sensor	24. EGR Solenoid
2. Fuel Pump	14. Oxygen Sensors*	25. Air Suction Control Solenoid*
3. Main Relay	15. Alternator	26. By-pass Solenoid B
4. To Starter	16. Cooling Fan Switch	27. By-pass Solenoid A
5. Cyl/Crank Sensor	17. Power Steering Switch	28. Air-Con Clutch Relay
6. TDC Sensor	18a. Neutral Switch (MT)	29. Check Engine Light
7. MAP Sensor	18b. A/T Position Switch	30. EAT ECU
8. Atmospheric Pressure Sensor	19. Vehicle Speed Sensor	31. Clutch Switch M/T
9. Throttle Angle Sensor	20. Injectors	32. Radiator Fan Control Unit
10. Ignition Timing Adjuster	21. Injector Resistors	33. Cruise Control
11. EGR Lift Sensor	22. E.I.C.V.	34. Igniter Unit
12. Water Temperature Sensor	23. Pressure Regulator Solenoid	

* Emission Vehicles Only

Fig. 6.46 A circuit diagram for a fuel injection system

EFI SYSTEM — Troubleshooting (Toyota)

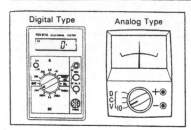

4. Use a volt/ohmmeter with high impedance (10 k Ω/V minimum) for troubleshooting of the electrical circuit. (See page FI-25)

TROUBLESHOOTING PROCEDURES

SYMPTOM — DIFFICULT TO START OR NO START (ENGINE WILL NOT CRANK OR CRANKS SLOWLY)

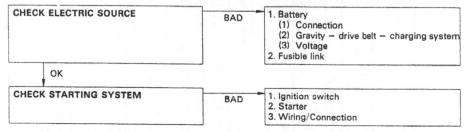

CHECK ELECTRIC SOURCE	BAD	1. Battery 　(1) Connection 　(2) Gravity — drive belt — charging system 　(3) Voltage 2. Fusible link

OK

CHECK STARTING SYSTEM	BAD	1. Ignition switch 2. Starter 3. Wiring/Connection

SYMPTOM — DIFFICULT TO START OR NO START (CRANKS OK)

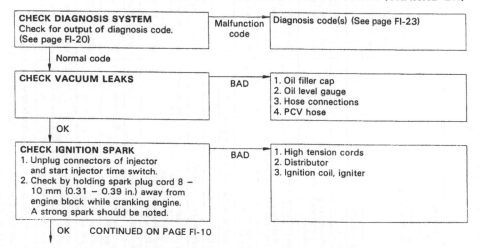

CHECK DIAGNOSIS SYSTEM Check for output of diagnosis code. (See page FI-20)	Malfunction code	Diagnosis code(s) (See page FI-23)

Normal code

CHECK VACUUM LEAKS	BAD	1. Oil filler cap 2. Oil level gauge 3. Hose connections 4. PCV hose

OK

CHECK IGNITION SPARK 1. Unplug connectors of injector 　and start injector time switch. 2. Check by holding spark plug cord 8 – 　10 mm (0.31 – 0.39 in.) away from 　engine block while cranking engine. 　A strong spark should be noted.	BAD	1. High tension cords 2. Distributor 3. Ignition coil, igniter

OK　CONTINUED ON PAGE FI-10

Fig. 6.47 The algorithm (decision flow chart)

2.7 DIAGNOSIS

NO. OF LED BLINKS BETWEEN TWO-SECOND PAUSES		SYMPTOM INDICATED	POSSIBLE CAUSE
0	LED ON	Engine will not start Engine runs at reduced power	Disconnect ECU ground wire Faulty ECU
	LED OFF	Engine will not start No particular symptom shown	Activated inertia switch Faulty ECU earth Disconnected ECU ground wire Short circuit in warning light wire Faulty ECU – Check system using 2.7 PGM-FI Fast Check or Cobest before changing ECU
1		Incorrect idle	Front oxygen sensor electrical fault
2		CO	Rear oxygen sensor electrical fault
3 5		Fuel fouled plug Frequent engine stalling Hesitation	MAP electrical fault Disconnected MAP pipe
4		Reduced performance	Crank angle electrical fault
6		High idle speed Difficult starting at low temperature	Coolant sensor electrical fault
7		Flat spots High idle speed	Throttle angle sensor incorrectly adjusted or electrical fault
8		No back-up facility and no cranking fixed timing	Distributor TDC sensor electrical fault
9		Reduced performance	No. 1 TDC cylinder sensor electrical fault
10		Slight effect on performance	Intake air temp. sensor electrical fault
11		No particular symptom shown except high idle speed	IMA electrical fault
12		Frequent engine stalling Erratic running at low speed No fault shown	EGR system electrical fault Sticking EGR solenoid or blocked pipe
13		Slight effect on performance	Atmospheric pressure sensor electrical fault
14		Idle speed hunting from 1,200 to 1,500 rev/min Erratic idle	E.I.C.V. electrical fault
15		Erratic running Engine will not start	Igniter unit electrical fault
17		No speedometer Stalling when vehicle comes to rest	Vehicle speed sensor electrical fault
18		Unable to adjust ignition timing	Ignition timing adjuster electrical fault

NOTE: 1. Always check wiring for open or short circuit before changing sensor.
2. Oxygen sensors are not fitted to all markets, if 2 blinks are shown this indicates a fault with the EXKPRO resistor fitted on the ECU in place of the oxygen sensors.
3. If code 16 or blinks exceeded 18, count number of blinks again, if the LED is still showing these codes, substitute a known good ECU and re-check.

Fig. 6.48 A fault finding (trouble shooting) chart

Summary

The material covered in this chapter gives an insight into the methods and equipment that are used for diagnosis. The examples chosen are indicative of the procedures used. That is to say, they serve as a guide to the procedures that can be deployed for fault diagnosis across a range of systems.

There is no intention to suggest that these are the only tools available to do the job. However, it is my belief that the basic approach, which is that the diagnostic tool will lead the operator to the general area of the fault, and after that it is a matter of conventional circuit testing, is best explained by concentrating on one of the simpler methods, as I have done here.

Various other types of diagnostic equipment are described in *Automobile Electrical and Electronic Systems* by Tom Denton, published by Arnold.

Early applications of electronics to vehicles resulted in different makes of systems being used: for example, an engine fuelling system of one make, a cruise control system of another make, and anti-lock brakes of yet another make. This resulted in separate diagnostic approaches for each system so that the test equipment for one system would not necessarily be suitable for use on other systems.

It is my impression that vehicle electronic systems are a bit like personal computers, in the sense that they are subject to constant developments. This is the case in the area of diagnostics and I can envisage the time when it will be possible to access all electronic systems on a vehicle through a single interface.

Developments in the USA, where a system known as OBD 2 (On Board Diagnostics stage 2) has obliged manufacturers to provide a standardised mode of access to a limited amount of diagnostic data. It is believed that a similar system will apply in Europe around the turn of the century and, eventually, it may be that diagnostics will become less of a problem. This topic is covered in Chapter 7.

In the introduction to this book I stated that some electronics knowledge is just one of the elements of skill required for work on electronic systems. Ability to use diagnostic equipment is another. I have also shown that comprehensive knowledge of vehicle technology and the way that elements of vehicle systems relate to one another is another important element of the skills set required for accurate electronic system diagnosis. It is also clear that sound knowledge of electrical principles and test methods, such as voltage and resistance checks, are essential skills for modern vehicle technicians.

Diagnostic equipment does not provide a complete solution; it is just another 'essential' element of the tools and skills set that will assist a technician to use a strategy, such as the 'six steps', to reach a solution.

Another important skill is the ability to examine a system visually, as a preliminary step to testing, in case there is some obvious problem like a loose connection.

The importance of understanding the inter-relationship of all parts of a system cannot be overstated. Recent experience (1996), when the new MOT emissions test was introduced, showed that an exhaust gas test failure was in some cases treated as a failed exhaust catalyst. Sometimes a new exhaust was fitted and a retest was performed, which produced another failure, and some consternation (to say the least). Had the individuals performing this test been aware that the catalytic converter is but one part of an emission control system, they should have known that other parts of the system were at fault, not the catalyst.

To summarise, whether the diagnostics is 'on-board' or 'off-board' it serves broadly the same purpose. That is, it directs the user to an area of a system (circuit) where a fault is indicated. Once that has been done the knowledge and skill of the technician is required in order to perform the further diagnosis that will lead to a solution of the problem.

7
Trends and developments

Key skills

In this chapter we discuss the consolidation of work in previous chapters and the justification for its inclusion in terms of technician activity.

Careful study over a period of time shows that the skill that leads to successful diagnostic work on vehicle electronic systems consists of many elements; the most important of these may be classed as 'key skills'.

These key skills may be summarised as:

1. Use appropriate 'dedicated' test equipment effectively.
2. Make suitable visual inspection (assessment) of the system under investigation.
3. Make effective use of wiring diagrams.
4. Use instruction manuals effectively.
5. Use multi-meters and other (non-dedicated) equipment effectively.
6. Interpret symptoms of defective operation of a system and, by suitable processes, trace the fault and its cause.
7. Work in a safe manner and avoid damage to sensitive electronic components.
8. Fit new units and make correct adjustments and calibrations.
9. Test the system and the vehicle for correctness of performance.

By examining a limited number of vehicle systems it has been possible to show what each of these 'key skills' comprises in terms of skill and knowledge.

To clarify the point the above list of nine key skills is taken, in numerical order. Each of these 'key skills' is then cross-referenced with topics covered in the text in order to identify the type of knowledge and skill involved in the procedures described. It should then be possible for readers to assess their own competence in the skills for dealing with electronically controlled vehicle systems, and to take action as required so that they can develop their skills to become competent at fault finding and maintenance tasks on these vehicle systems.

Key skill 1: Use appropriate 'dedicated' test equipment effectively.

In Chapter 6 the use of two types of 'dedicated' test equipment was examined in some detail. One was on-board: that is to say, the fault codes are read out from the ECU 'blink code' system and then interpreted with the aid of a table. The other system used off-board equipment; here the fault codes are converted into text, which makes the job a little easier.

The main factors to be noted are that the diagnostic systems examined are typical of the types used in the vehicle repair industry. In both the on-board and the off-board cases, the equipment is designed to direct the user to a sub-system (circuit) that is defective.

The equipment is reasonably easy to use but, because it is dedicated to a specific vehicle model, it is imperative for the user to be familiar with that model, or to have to hand manuals and other data that give the instructions for operating the system.

The bulk of the electronics and computing is performed by the ECU and the diagnostic equipment operates through the ECU. This means that most of the electronics testing is done by the diagnostic equipment. The operator does not require a high level of electronics knowledge to obtain quality output from the diagnostic equipment.

When the diagnostic output is obtained, it is in the form of a code which may require decoding. In the case of on-board 'blink' codes, the decoding takes the form of a chart. When the diagnostic output is obtained from off-board equipment, of the type shown in Chapter 6, it is usually in text form.

It is also important to note that the equipment and procedures for diagnosis of heavy vehicle electronic systems is much the same.

Key skill 2: Make suitable visual inspection (assessment) of the system under investigation.

Earlier chapters make the point that an informed visual examination of the suspect system may reveal an 'obvious' fault, for example a perforated vacuum sensor pipe or a wire disconnected from a coolant sensor. Early detection of such defects can save much time.

To perform an informed visual inspection it is necessary to know what one is looking for. Chapters 1 to 5 emphasise the basically common structure of electronically controlled systems and show that they consist of sensors, actuators, interconnecting cables and an ECU. This is held to be an important concept because it means that when one is attempting to become familiar with a system it is possible to approach the learning process in a structured way.

To understand what a vehicle electronic system consists of, so that garage type inspections can be made, it is necessary to know what functions the system is intended to perform in terms of vehicle operation. The examples given show the type of understanding that is necessary in the case of ABS, fuel injection, ignition and charging systems.

These descriptions are based on the assumption that the reader has a sound knowledge of conventional vehicle technology, e.g. how engines work, how brakes work, etc. Without this understanding of the mechanical engineering technology of vehicles it is difficult to see how a technician will be able to conduct an informed visual inspection of a vehicle's system. It is helpful to know that the 'electronics' are there to help the mechanical parts work more efficiently, in terms of performance, or legal requirements such as emission levels, and trailer anti-skid legislation.

The location charts, which appear several times, are a valuable aid to finding the location of a component on the vehicle.

Key skill 3: Make effective use of wiring diagrams.

Chapter 3 introduced some electrical revision and emphasised that knowledge of basic electricity and circuits is an absolute necessity. The material shown is intended as an aid to memory. If the reader does not have the basic knowledge indicated then it is essential that they acquire it because it is fundamental to the proper execution of diagnostic and repair work on electronically controlled vehicle systems.

It is not the intention to provide a textbook on basic electricity. However, the point is made that vehicle technicians may find it easier to acquire this basic electrical knowledge by following a course which concentrates on the building of practical circuits, where inputs and outputs may be observed and measured, rather than a more conventional course where the algebra associated with Ohm's law is often found off-putting.

The diagnostic equipment, whether on-board or off-board, directs the user to a circuit. Testing in that circuit requires a knowledge of the circuit structure. The nature of this structure is contained in the circuit (wiring) diagram.

Several circuit diagrams are shown. They range in complexity from the full vehicle wiring diagram which, in the case of Rover, has a grid system to facilitate location of an area of the circuit, to the relatively simple diagram of a fuel pump circuit. The symbols associated with wiring diagrams are shown and the 'danger' of assuming that only standard symbols are used is highlighted.

The colour coding system is also explained because cable colours are of great assistance when tracing wires.

Key skill 4: Use instruction manuals effectively.

The reader has been constantly reminded of the critical importance of having available all the information relating to the system being diagnosed. Using the manuals in a constructive way is an important skill, and the need to have a secure place on which to rest the manual while working on the vehicle cannot be overstated.

A careful study of the relevant sections of the manual, prior to starting diagnosis, is time well spent. There is an often-repeated saying: 'When all else fails, read the instruction book.'

Reading the instruction book prior to starting work on the system is an essential part of the skill, certainly for those who are not familiar with a given vehicle system. A later section shows modern systems that use CD ROMs to hold the instruction manual so that it can be accessed through the screen of the diagnostic computer.

Key skill 5: Use multi-meters and other (non-dedicated) equipment effectively.

Much of the material in Chapters 4 and 5 shows that many sensors and actuators rely on electro-mechanical principles for their operation. The performance/condition of these devices can be assessed by voltage and/or resistance measurements. This requires the ability to make these measurements to the required degree of accuracy. A good quality moving coil instrument with an impedance value of 20 000 ohms per volt is satisfactory. However, a digital type meter is preferred.

Part of this skill requires that the instrument should be checked for accuracy at regular intervals. The leads, clips and probes need to be kept in good condition, and multi-meter batteries must be up to voltage. In most cases, test lamps are not recommended because the high current involved may damage sensitive parts of a circuit.

Key skill 6: Interpret symptoms of defective operation of a system and, by suitable processes, trace the fault and its cause.

This covers a wide area. The problems of assuming that a failure to pass the emissions part of the MOT test is caused by a failed catalyst have been stated. The fact that the failure may be caused by a blocked air filter is but one example. The early chapters contain constant reminders that an electronically controlled system can only work satisfactorily when all inputs and all outputs are correct.

In the section on ABS, the need to know that a wheel bearing failure, or incorrect adjustment of the bearing, is a common cause of reported sensor failure, is another indication of the level and type of expertise that a vehicle technician needs.

Key skill 7: Work in a safe manner and avoid damage to sensitive electronic components.

The importance, of working systematically is emphasised and the 'six step' approach is recommended for this purpose. If technicians do not work in this way they are likely to run into problems. It is a good, disciplined way to work on vehicle electronic systems and it has useful application in many other tasks.

The elements of the six steps can be seen in the description of the use of diagnostic equipment in Chapter 6, where 'collecting evidence' is prominent.

Warnings of the dangers associated with high voltages from ignition systems and the risk of injury involved in other operations is emphasised.

The possibility of damage from static electricity is described, and the necessary precautions to be taken when welding on a vehicle are stated. In this connection the publication of The Institute of Road Transport Engineers' *Code of Practice for the Installation, Maintenance and Repair of Electronic Systems Fitted to Commercial Vehicles* is recommended as a valuable guide to the type of damage prevention measures that are recommended by vehicle and equipment manufacturers.

Key skill 8: Fit new units and make correct adjustments and calibrations.

The point is made that most 'electronic' repairs on vehicles require replacement of a defective part. This is sometimes taken quite literally and often leads to individuals attempting to make a repair by an almost random approach where each element of a system is replaced in the hope that it may 'cure' the problem.

The fact that a failed component may have been caused by a defect external to the component is often overlooked. This results in the newly replaced unit being damaged. Once again, the 'six step' approach comes into play. The cause of the failure must be found and repaired before a new unit is fitted. That is the beginning of the story. Fitting a new unit may be simply disconnecting a few wires, undoing a few screws, removing the old unit and then reversing the procedure when fitting the new unit. However, in many cases, it is not that simple.

Very often there is a good deal of vehicle engineering skill required. Take the case of the oxygen sensor shown in Figure 4.11(b). This particular sensor is reasonably accessible. However, it is in the hottest region of the exhaust pipe, and it will be necessary to allow the engine to cool before commencing work. When working on any part of an electrical system where wires are to be disconnected it is advisable to disconnect the battery at its earth terminal.

The sensor may have become 'seized' in the screw thread and the necessary steps will need to be taken to overcome this difficulty. It should be noted that the sensor is quite close to the vehicle's radiator, so care must be taken to ensure that spanners do not slip and cause damage to the radiator. There are probably other factors that relate to this particular operation, but this short description should serve to show that considerable skill is required in the apparently 'simple' task of fitting replacement units.

Key skill 9: Test the system and the vehicle for correctness of performance.

This is one of the six steps; it is one that is sometimes overlooked.

Should it be an intermittent fault that has been rectified an extended road test may be required in order to establish that the repair has been effective.

In some cases adjustments to stepper motor linkages, and solenoid operated valves, require that the completed adjustment must be checked to ensure that the actuator functions correctly. This check may be performed before the system itself is put into operation.

I think I have now said enough to show some of the details of technician skill that will lead to successful work on electronically controlled vehicle systems. I hope that I have not trivialised the subject but at the same time I also hope that I have provided enough information to show that the skills are perfectly within the reach of most people, many of whom it has been my privilege to teach over a period of many years.

Most of the additional skill that is required is electrical: testing circuits for resistance, testing sensor voltage, checking ECU outputs and actuator inputs. All of these activities have been described and shown to be vitally important.

I have restricted the number of systems described to the minimum consistent with my stated purpose. That purpose is to show that most electronically controlled vehicle systems have a common structure. This basic concept, in my view, enables one to place electronics in context and to decide on a learning process that will lead to successful diagnostic and repair work on modern vehicles.

The examples of systems discussed are widely used. Any reader wishing to study vehicle electronic systems in greater depth will find that the publication *Automobile Electrical & Electronic Systems* by Tom Denton, published by Edward Arnold, covers a wide range of vehicle systems.

Access to repair information

Independent garages, i.e. those not tied to a particular vehicle manufacturer, sometimes have difficulty performing repairs on electronically controlled systems because they do not have access to the necessary technical information. A remedy for this problem is for the independent repairer to direct the vehicle owner to an authorised dealership for the repair to be performed there. Some argue that this is not a satisfactory state of affairs and that technical information, necessary for repairs to be effected (not alterations), should be available to all bona fide vehicle repairers.

An apparent benefit of freely available repair information is that a single set of diagnostic equipment could be designed so that, when placed in the hands of a bona fide vehicle repairer, any electronically controlled system on any make of vehicle could be diagnosed. I make no further comment on the issue, except to say that certain European legislation regarding access to information relating to emission control systems may mean that more technical information will be made available to independent repairers around the year 2000. This seems to be a rational development.

On-board diagnostics (OBD)

In the USA there is a requirement for on-board diagnostics to provide 'layman's' information about the operational state of electronically controlled emissions systems. The first stage of the USA legislation was called 'On-board-diagnostics 1', and the second stage, which became operational around 1994, is known as OBD2. (By 'layman', I mean a properly trained and competent technician, as opposed to an electronic designer.)

The UK authorities have been active in drafting proposals for a similar OBD requirement for European vehicles. In general terms the European OBD would require a vehicle to have two warning lights, one amber and one red. The red light would be used for safety related problems in the emission control system, and the amber light to advise the driver of malfunctions that were affecting proper functioning of the emission control system. These warning light signals would be associated with a set of maintenance information that would permit a competent technician to make the necessary adjustments and repairs to bring the system back to operational efficiency.

Movement towards a standardised approach to diagnostics for the vehicle industry has obvious attractions for many involved in vehicle repair and break-down services.

Networking of ECUs

When a vehicle is equipped with a number of discrete electronic systems, such as engine management, ABS, cruise control, etc., it is to be expected that each system will have its own ECU. These ECUs can be accessed individually for diagnostic purposes.

As systems developed it became evident that it was necessary for ECUs to communicate with each other in order to make systems such as traction control and transmission control work effectively. For example, if ABS is used as traction control, to stop wheel spin it will be necessary to momentarily ease engine power. For this to happen there must be 'real time' communication between the ECUs. The speed of communication between ECUs may be of the order of a million bits per second (M bits/second). This leads to the need for a high speed data bus. We thus have a high speed data bus to add to the slower speed data bus that is used for multiplexed wiring as described in Chapter 8.

Networking (linking) of discrete systems and their ECUs can lead to savings. For example, the engine speed sensor that is used by the engine controller can also be used by the transmission controller. However, because the systems are now linked and the operation of one system is dependent on the state of the linked systems, diagnostic information must be drawn from the network through a common diagnostic interface, as shown in Figure 7.1.

Data buses and their design are covered by a set of 'rules' that have become known as CAN (Controller Area Network). CAN affects vehicle repair in the sense that it makes approaches to repair and diagnosis somewhat different.

As the processing power of micro-controllers increases, the concept of a single computer to control all systems becomes progressively more realistic. This would probably make the job of diagnostics easier.

When this type of networking is built into a vehicle the need for electronic access to repair data becomes increasingly important because it is no longer feasible to use multi-meter type, intrusive testing, certainly not at the same level as with earlier non-multiplexed systems. It thus becomes necessary to incorporate facilities for access to diagnostic information. Developments in this area have been underway for some time and they fall under the heading of 'Keyword

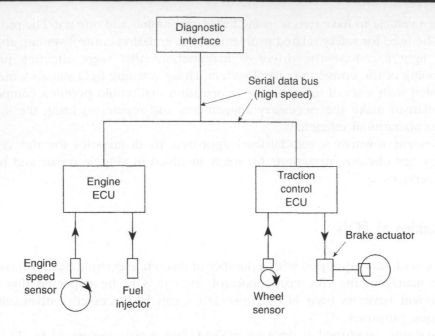

Fig. 7.1 Diagnostic interface for networked systems

Protocol 2000', KWP 2000, which is a comprehensive set of proposals for a standardised approach to the design, operation and diagnostics of multiplexed and networked systems.

As more complex systems are introduced into common use, we may expect to see further advances in both on-board and off-board diagnostics. Once again, the need for good quality, up-to-date and relevant training is reinforced.

Standardised fault codes

As stated in the section on logic circuits, in Chapter 8, the data that is passed around the controller is 0s and 1s, and these are combined to make various codes (words).

When these 0s and 1s are transmitted, they do so 'serially', i.e. 'one bit at a time'. This is referred to as 'serial data communication'.

The technology for reading serial data is readily available but, as it is coded, the problem remains 'What does the code mean?'.

There is an SAE (Society of Automotive Engineers) standard for fault codes on American trucks. This standard (Ref SAE 1587) advocates the use of common codes applicable to all vehicles. Computer codes are complicated, but just to illustrate my point about the use of common codes I will give an example.

A fault code as read at the diagnostic interface may be made up of two 'computer words' – bytes. A byte is 8 bits, i.e. 0s and 1s. The failure mode identifier part of 1 byte is 4 bits long. The SAE 1587 code for 'bad component' is

12 (ordinary decimal); in binary this is 1100. Other parts of the code relate to system, e.g. engine. Such a fault code system that always used the same code for 'bad component' when reporting on a given system would seem to offer attractions in terms of simplifying procedures.

The extent of movement in this direction is difficult to assess. However, at a seminar in London (December 1996) I noted that one UK manufacturer described an electro-pneumatic braking system for heavy vehicles that incorporated SAE J1587 and also SAE J1939. The 1587 refers to standardised fault codes and their communication, and the 1939 makes recommendations about other factors relating to the design and operation of heavy vehicle systems.

Interference

As the amount of serial data (0s and 1s) in the form of clocked voltage pulses (low, high) transmitted around the vehicle data bus increases so other problems, such as electro-magnetic (inductive) and capacitive interference between cables, may be created. When an electrical pulse in one cable interferes with current being transmitted through a nearby cable the effect is known as 'cross talk', or interference. Most readers will be familiar with radio interference and will know that it is considered good practice to fit the aerial as far as possible away from the ignition system, e.g. on the roof, as shown in Figure 7.2(a).

With cables that are conducting data pulses it is impracticable to separate them by a distance that would prevent 'cross talk', so other measures are adopted. Some of these measures are discussed below.

Fig. 7.2(a)

COUNTERING 'CROSS TALK'

To overcome problems such as 'cross talk' (one signal interfering with another) various measures are employed: for example, screened cables and 'twisted pairs'.

Twisted pair

Figure 7.2(b) shows the twisted pair of cables. Capacitive interference from a nearby cable is reduced because the two wires are kept close together and this helps to make the signal-carrying wires less susceptible to capacitive 'cross talk'. The twisted pair is also used to counter inductive interference from nearby cables. The theory is that the 'interference' effects cancel one another out over the length of the cable.

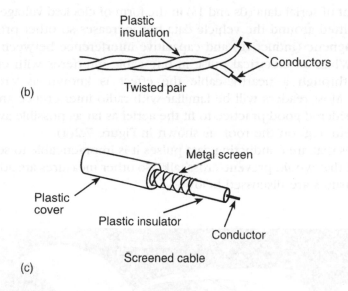

Fig. 7.2(b) and (c) Twisted pair – and screened cable

Screened cable

Figure 7.2(c) shows the screened cable (co-axial). Here the interference is reduced because the interfering current is mainly confined to the outer surface of the screen where it does not distort the signal in the central conductor. The earthing of the screen is an important factor.

These features are mainly design factors, but repairers need to be aware of them so that problems are not caused by fitting incorrect cables, or removing screening.

Figure 7.3 shows a Crypton GM circuit diagram where shielded cables are used on items 7 and 8, the engine speed sensor, and the exhaust oxygen sensor.

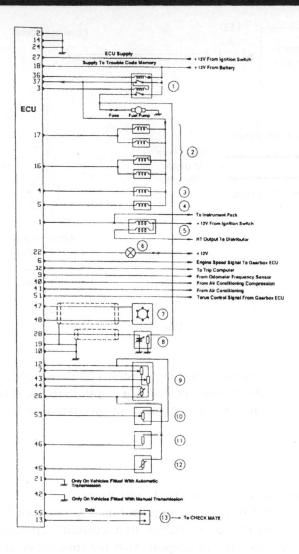

Fig. 7.3 A circuit using screened cables

Fibre optics

Fibre optics operate, very broadly, on the principle described here. Figure 7.4 shows a simplified arrangement where an LED converts electrical pulses into light energy. This light energy passes through the transparent glass fibre. At the receiving end of the cable an opto-electronic device, in this case an opto transistor, converts the light back to an electric current. The advantage of this method of signal transmission is that fibre optic cables do not give out any electrical interference.

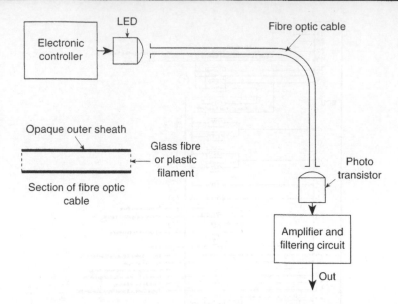

Fig. 7.4

In order to further improve matters, systems that use fibre optics for data transmission are being developed because fibre optics are not subject to the same electro-magnetic problems as current pulses in ordinary cables.

Self-diagnostics

As systems become more sophisticated, and more of the vehicle's operational functions are committed to computer control, so the need for better diagnostics grows. The material in this book shows that current 'self-diagnostics' (on the vehicle) are fairly limited and this suggests that the time when the computer will replace the technician is far distant. However, that is not to say that diagnostic systems cannot be improved. One has only to look at the way things have changed over the past 10 years to appreciate that even more significant progress can be made. To assess this aspect of vehicle diagnostics it is worth considering a commonly used device – the home computer.

Home computers have a certain amount of resident (self) diagnostics. My own machine often reports a printer fault when I attempt to print something. This usually happens because I have not loaded the paper. However, it could be any one of a large number of faults. But the resident diagnostics (processor) only reads certain values; it is not programmed to search the printer circuit for faults. This highlights the sort of problem that occurs with vehicle diagnostics; the diagnostic information is very brief.

As explained in Chapter 6, the current range of diagnostic tools direct the user to a circuit, or general area, of a system. However, as more processing power becomes available it is possible to envisage the development of more detailed diagnostics by sampling a greater number of variables (data values) at selected points in the system.

It is then a matter of programming the micro-controller to carry out routines that will perform analytical assessments of the circuit data. These assessments would probably be based on the physical laws that govern the operation of the circuit. For example, if a sensor voltage were to be taken at the sensor and then at the controller end of the sensor cable the two could be compared, by the processor, and if found to be out of limits a fault code could be generated, for that particular case. Of course, this assumes that the expense resulting from extra data inputs could be justified. However, because the pace of change in micro-electronics is rapid, very little can be ruled out.

Handling information

The earlier chapters of this book have made frequent references to wiring diagrams, location charts, instruction manuals and other aids to assist with repair work on vehicles. Handling the documents and the information contained in them is a major factor in repair work and any development that will make this aspect of the work more manageable is to be welcomed.

An estimate relating to the USA vehicle repair industry puts the number of manuals required to repair all makes of vehicle at 100 for the year 1995. This is predicted to grow to 150 by the year 2000. The number of pages is estimated at over 200 000 and it is suggested that some 60 or so of these manuals will be available in 'electronic' form. I do not know the extent of the problem in the UK, but earlier research showed that a significant number of technicians experienced difficulty in making use of manuals and the updates that follow modifications to vehicles and changes in their specification.

However, information technology progresses at a rapid rate and it is to be expected that UK companies will wish to be at the forefront of developments. There is some evidence that a trend towards computerisation of diagnostics is taking place in the UK, and to assess the effect of this at the garage industry interface I interviewed a number of managers, one of whom invited me to a demonstration of the computerised equipment that is used in his Rover dealership. The following description of the equipment is based on this fact-finding exercise.

The equipment shown in Figures 7.5(a) and (b) is Rover Group 'TESTBOOK'. It is based on a laptop computer of a late specification.

The base unit incorporates a dual CD ROM drive. A range of adaptor kits, to suit all vehicles in the product range, is contained in the plastic cases shown in Figure 7.6 (the larger ones contain the adaptor kits for the earlier versions of electronic controls).

(a)

(b)

Fig. 7.5(a) and 7.5(b) 'TESTBOOK' diagnostic equipment

Figure 7.5(c) shows the protective cover which is placed over the equipment when not in use. A Rover 400 series vehicle was used for the test demonstration.

Figure 7.7 shows the set of adaptor cables that are used for testing individual systems on the vehicle, such as engine management, anti-lock brakes, etc. All systems on the vehicle are accessed through a common diagnostic plug by the use of these adaptors.

The diagnostic (serial data connection) connector is situated under the dashboard, on the driver's side, and Figure 7.8 shows it being held in the technician's

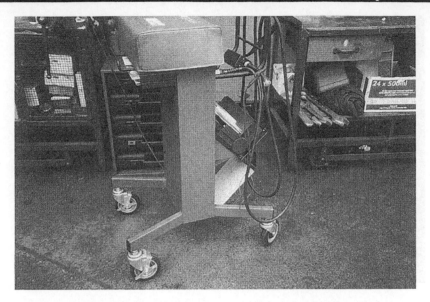

Fig. 7.5(c) Protective cover over equipment when not in use

Fig. 7.6 Adaptor kits

hand. The TESTBOOK machine is connected to this connector via the adaptor cable for the system under test.

When the vehicle and test equipment have been prepared the CDs are placed into their respective drives. There are two CD drives and two CDs for each vehicle

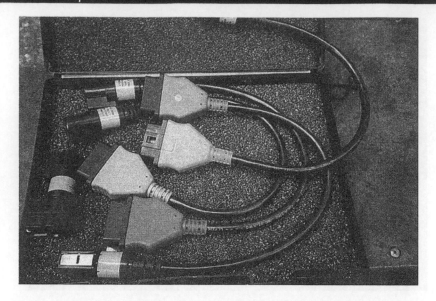

Fig. 7.7 Adaptor cables for all systems on the vehicle

Fig. 7.8

model. One of these CDs carries the test program and the other the diagnostic information, as shown in Figure 7.9.

The test program guides the technician through the test sequence results and messages appear on the computer screen in response to actions taken.

Fig. 7.9 The two CDs. One for the test program, the other for diagnostic information

The diagnostic CD contains location charts, pictures showing exactly where, on the vehicle, a particular sensor or other device is located, and recommended procedures for dealing with a problem; in fact, practically everything that would be found in the normal 'paper and print' type of workshop manual.

When the test commences the 'touch' screen of the computer displays icons. Touching a particular icon will direct the machine to start the required test.

If a fault is registered instructions are issued, through the screen, to guide the technician to the next step. As with other diagnostic equipment the serial data codes are restricted to a fairly high level of fault reporting. The printout shown in Figure 7.10 gives an indication of this. For example, 'Throttle Switch Circuit': should this be reported as 'fault' further testing within that circuit will be required. This testing will often consist of the types of voltage, resistance and current tests that are described in this book.

However, with this equipment it is not necessary to be equipped with separate meters because they are contained 'in the computer'. When a voltmeter is required the 'icon' is touched and a voltmeter is displayed on the screen. Figure 7.11 shows the various icons. Figure 7.12 shows the technician holding the voltmeter probes and the connector.

As with other tests, the screen information will guide the user to the 'pins' or sockets (cavities) where the test probes are to be applied. It is evident that this type of equipment has advantages over previous types; for example, the easy access to test programs, the convenience of having workshop manual information on a particular topic, available on screen, the built-in meters, etc. all point to a diagnostic system that puts computer technology to work in a very constructive way. However, lest one is tempted to think that technician skill has been made redundant, I think that it is worth considering a few details.

```
SERIAL COMMS _____ GOOD
BATTERY VOLTAGE _____ GOOD
VEHICLE ECU IDENT _____ GOOD
COOLANT SENSOR CIRCUIT _____ GOOD
COOLANT OPERATOR _____ GOOD
AMBIENT SENSOR CIRCUIT _____ GOOD
AMBIENT OPERATOR _____ GOOD
THROTTLE SWITCH CIRCUIT _____ GOOD
STEPPER MOTOR _____ GOOD
IGNITION PULSE INPUT_____ GOOD
IGNITION SWITCH _____ GOOD
STARTER MOTOR CIRCUIT _____ GOOD
OVERRUN FUEL CUT OFF SOLENOID _____ GOOD
TEMPERATURE GAUGE _____ GOOD


TUNE COMPLETE


CO SET TO 3% + OR – 0.5%
FAST IDLE SPEED = 1164 RPM
BASE IDLE SPEED = 771 RPM
```

Fig. 7.10

The machine cannot think for itself. As explained in earlier chapters, failure of a vehicle to comply with emission regulations may not be due to the exhaust catalyst or to a malfunction of the electronically controlled fuelling system; it may only be due to a relatively simple mechanical defect. Suppose that there is a tight (no clearance) valve clearance which is causing a loss of compression: there will not be much point in connecting up for a diagnostic test of the electronic system. This is a situation that calls for vehicle knowledge.

In the first instance, a perceptive technician might detect an unevenness about the engine's running. A road test might reveal, when pulling hard at slow speed, the slight loss of propulsion that occurs when the 'weak' cylinder is doing the work. In general, the rule should be: carry out a thorough 'health check' on the suspect system before attempting to use the diagnostic equipment. This is one of the reasons why throughout this book emphasis has been placed on the extent of 'traditional' technology that exists in electronically controlled systems.

In Chapter 1 I stated that much of a vehicle technician's work consists of 'conventional' tasks such as repairing brakes, fitting oil filters, tuning engines, etc. While I was talking to my friend about the TESTBOOK equipment I noted some of the other jobs that were in progress in the workshop. They included fitting a new clutch, curing a miss-fire on a high mileage, 10-year-old vehicle. I believe this tends to justify the line I have taken. It also reinforces the view that good computer literacy is an essential ingredient of a modern vehicle technician's bank of knowledge, which must be added to the conventional 'mechanical' skills.

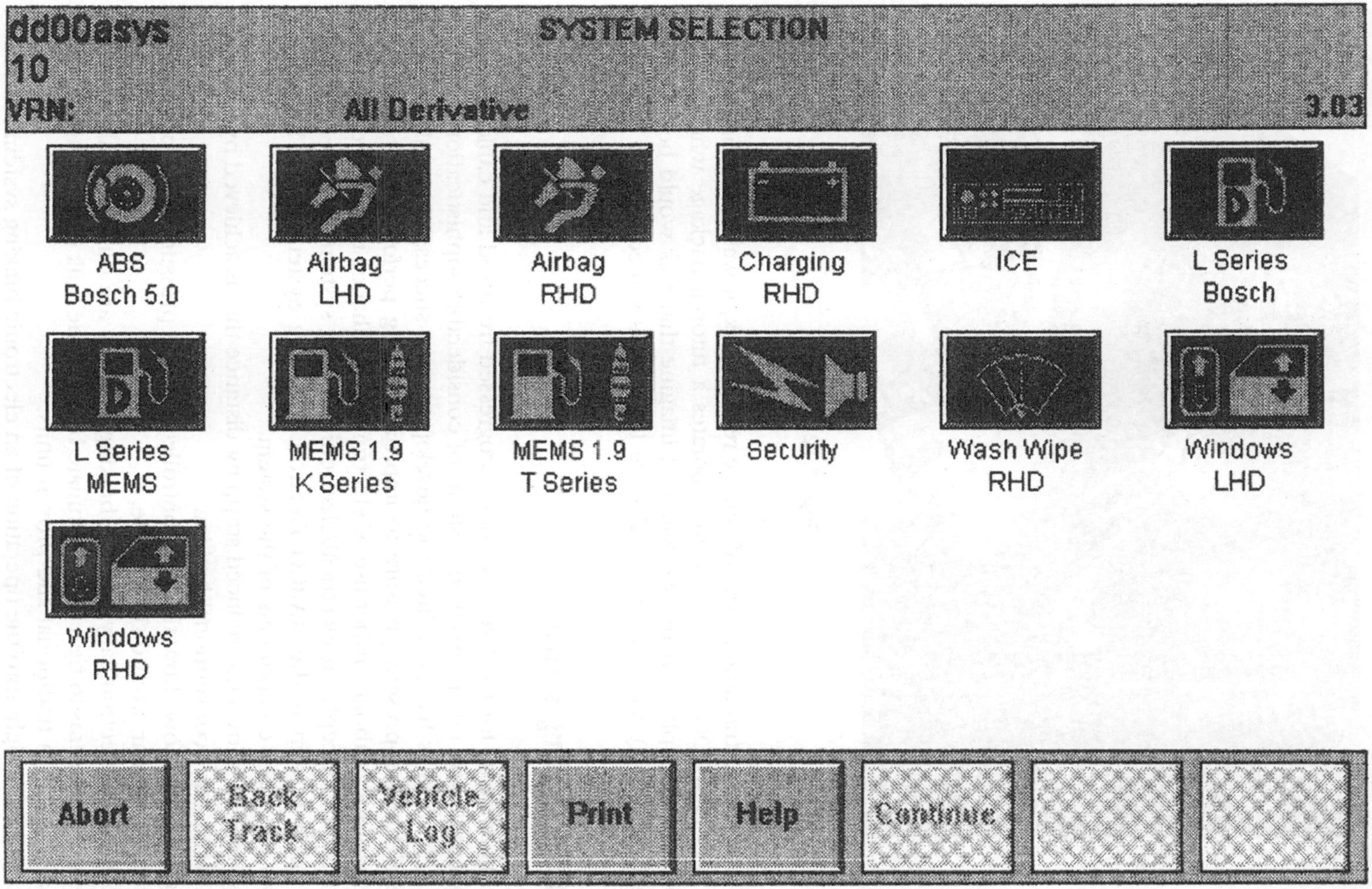

Fig 7.11

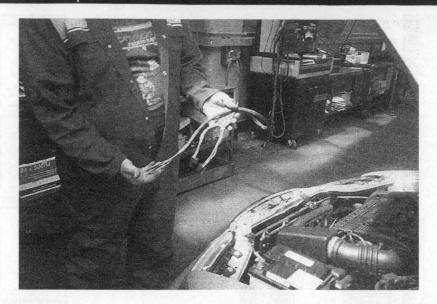

Fig. 7.12

While on the 'training' aspect of vehicle repair work, it was interesting to know that the 'TESTBOOK' equipment incorporates a tutorial package which enables training to take place in the workshop. I imagine that this would be a valuable asset in the NVQ setup, where workplace learning is valued.

Electronic braking system

Braking systems for road vehicles have progressed from rod and cable operation to hydraulic and compressed air systems of considerable sophistication. In recent years electronics has permitted the development of successful anti-skid systems (ABS). The ever-present pressure to improve braking performance has led to a variety of ideas about further use of electronics to enhance braking performance, and several concepts for electrical/electronic braking systems have been mooted. One such system is the WABCO Electronic Braking System (EBS). Figure 7.13 shows the main components of the system.

Figure 7.14 shows the reduced stopping distance that is achieved by EBS when compared with conventional braking.

Figure 7.15 shows how the EBS controlled braking pressure leads to more even brake lining wear around the vehicle.

Sometimes electronically assisted braking systems are dubbed 'brake by wire' systems. This seems to be a bit unfortunate because the bulk of the braking system is made up from traditional and proven units.

To give an insight into the operation of an electronic braking system I shall give a simplified description of the principle. This description does not relate specifically to any make of braking system.

WABCO EBS

1 = EBS electronic control unit
2 = Brake transmitter
3 = Proportional relay valve
4 = ABS solenoid valve
5 = ABS / EBS connection (ISO 7638)
6 = Trailer control valve
7 = Axle modulator
8 = Back-up valve
9 = EBS relay emergency valve
10 = EBS trailer modulator
11 = Wheel speed - and wear sensors
12 = Speed sensors

Fig. 7.13   WABCO EBS

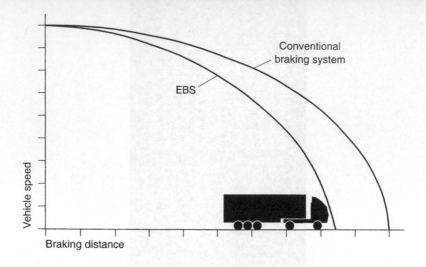

Fig. 7.14 EBS braking distance

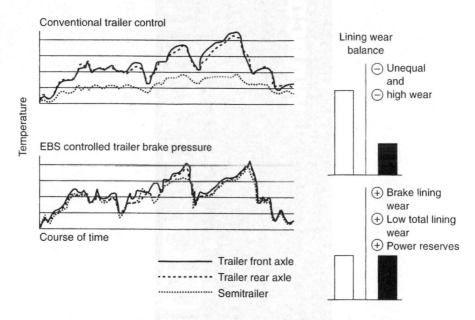

Fig. 7.15 Comparison of brake lining wear

The system I have in mind has a single circuit for the electronic control while retaining full compressed air (manual) operation as a fail-safe back-up. In this simplified system the driver's brake pedal valve incorporates a sensor which produces a voltage proportional to the required braking effort. This is achieved by the use of a potentiometer. The electrical signal from this sensor is transmitted to

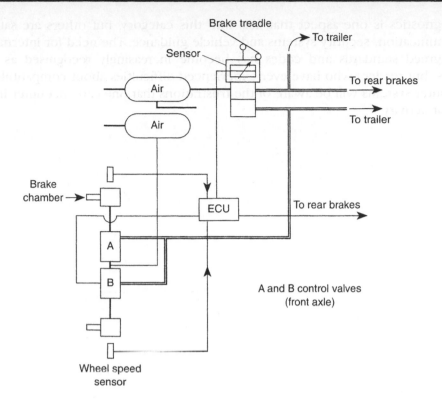

Fig. 7.16 A simplified version of electronic braking (front axle only)

the ECU which then sends a signal to the electro-pneumatic control valves which permit application of the brakes. The difference in time between electronic application of the brakes and compressed air application accounts for a reduction in stopping distance.

This is a much simplified version, but it serves to show how technology is moving. It also reinforces earlier work which shows that systems contain sensors, actuators, circuits and an ECU.

Summary

It will be appreciated that multiplexing and data communication are complex subjects. As the number of electronically controlled systems on the vehicle increases so the need to communicate between these systems increases. A branch of industry which is devoted to data transmission on the vehicle, and between the vehicle and the outside world, has evolved, an area of activity which is expected to grow.

Diagnostics is one aspect that falls into this category, but others are satellite communication, security systems and vehicle guidance. The need for internationally agreed standards and codes has become increasingly recognised as time passes, but readers who have ever experienced difficulties about compatibility of computer systems will be aware of the frustrations that one can encounter in this area of activity.

8
Underpinning knowledge

It is recognised that a good understanding of what makes systems work – and by extension what may be wrong when they do not work – is essential background education for vehicle technicians. This essential background knowledge is known as 'underpinning knowledge'. A good guide to the background knowledge that is required for work on vehicle electronic systems is to be found in the City & Guilds of London syllabus 383–50. This syllabus is entitled 'Competence in servicing, adjustment and checking of vehicle electronics; repair by replacement or modification'. The material presented in this chapter is based on this syllabus. Not every item in the syllabus is covered however, only those which the author feels will be most beneficial to readers of this book and its stated purpose.

Throughout the previous chapters attempts have been made to introduce electronic ideas in the context of the system to which those electronic principles (ideas) are applied. In this chapter a more conventional approach is adopted and the electronic devices are described as 'stand alone' items rather than elements of a complete system. However, where it aids the explanation, the motor vehicle circuit of which a particular device forms a part, a complete circuit is included.

Electronic components (discrete)

The term 'discrete' is applied to electronic components such as transistors, diodes, resistors, etc. Discretes are assembled on boards such as printed circuit boards (PCB) and veroboard to build the circuit required. Figure 8.1 shows a PCB as used in an early version of an ECU. On this PCB there are discrete components such as resistors and capacitors, as well as a number of integrated circuits (I/Cs).

RESISTORS

Resistors are widely used in electrical/electronic systems. For our purposes they may be divided into two categories:

Fig. 8.1   An electronic control unit PCB

- fixed value resistors
- variable resistors

Fixed value resistors are made from various materials. Among those commonly used are wire wound, moulded carbon and carbon film.

Wire wound

As the name suggests, a wire wound resistor is formed from wire which is wrapped around an insulating core. They are used in applications where power (heat) is dissipated in a relatively small space, and they are usually better able to tolerate the high temperatures caused by their operation than other types. They can also be made to a closer tolerance than some other types. For example, the rated value, say 20 ohms, can be accurate to ±5%.

Moulded carbon

Moulded carbon resistors are made from a paste of graphite and resin which is thoroughly mixed together, moulded into shape and then baked. The final product is then covered by a layer of insulating material. Very high resistance values are obtained.

Carbon film

These resistors are made by evaporating a film (layer) of carbon onto a ceramic core of insulating material. They have a low temperature co-efficient (temperature has less effect on their resistance value than other types), and they can be made to very close tolerances.

Colour coding of resistors

Resistors are colour coded by having coloured bands placed around their circumference, as shown in Figure 8.2.

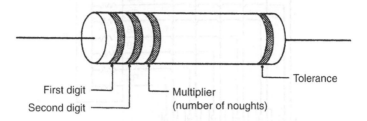

Fig. 8.2   Colour coding of a resistor

Different colours are used to represent the resistance value and its tolerance. With a four-banded code the first two bands give the significant figures, the third band is the multiplier, e.g. 1000, 100 000, etc., and the fourth band gives the tolerance, e.g 2%, 10%.

As an example, a brown first band means 1, a black second band means 0, a red third band means a multiplier of 100, and a silver fourth band means a tolerance of 10%. This means that the resistor has a value of 1000 ohms (1 kohm), ±10%.

A fixed resistor application

There are many uses of resistors on vehicles and the fuel injector circuit in Figure 8.3 shows one application that is relevant to electronic systems. This circuit is covered in Chapter 4 but it is included here also so that the resistor bank part of the circuit can be seen in greater detail.

In this circuit each resistor has a value of 5 to 7 ohms. If the value is outside these limits the circuit will not function correctly and the defective resistor must be replaced. Fig 8.3(b) shows the procedure for checking resistance values. Each resistor is checked individually, i.e multi-meter set to ohms and connected to (a) and (g), (a) and (f), and so on, until each has been checked.

It should be noted that the manual will normally show how to check the injector circuits by indicating the appropriately numbered contact points at the ECU connector. The resistor bank would be checked separately.

Variable resistors

The principle of the variable resistor is shown in Chapter 4, where its use as a throttle position sensor is discussed. Figure 8.4 shows a rheostat being used to control the intensity of illumination of the instrument panel lighting. A suitable

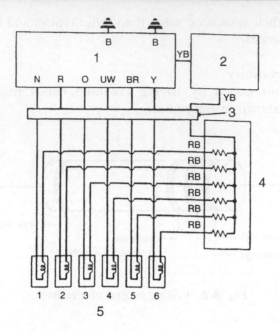

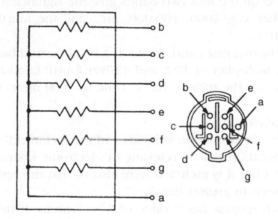

1. ECU
2. Main relay
3. Harness block connector R.H. suspension tower
4. Resistor pack
5. Injectors

Fig. 8.3(a) Resistors in petrol injector circuit

handle (knob) is attached to the spindle of the rheostat, and rotation of this handle gives more or less resistance in the circuit and hence more or less illumination, as required.

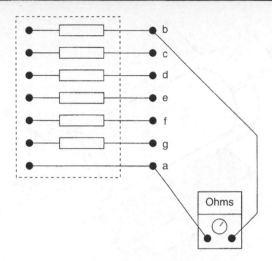

Fig. 8.3(b) Checking resistance values

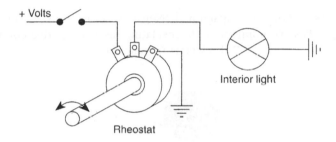

Fig. 8.4 A rheostat to control illumination

CAPACITORS

Capacitors (condensers) are made from two layers of electrical conducting material which are separated by a layer of insulation. The insulating layer is also known as the dielectric.

The dielectric may be a flexible material such as polythene, or paper treated with electrolyte, and the resulting 'sandwich' of insulation with a thin conducting (metal) foil on each side can be made into a roll. This roll can then be inserted into a canister, which gives a capacitor of cylindrical shape as shown in Figure 8.5.

Other types of capacitors

Mica and ceramic capacitors are also commonly used in circuits. In integrated circuits capacitors may be formed from a metallised layer placed on top of the film of silicon dioxide (the dielectric) on top of the silicon substrate, as shown in Figure 8.6(b).

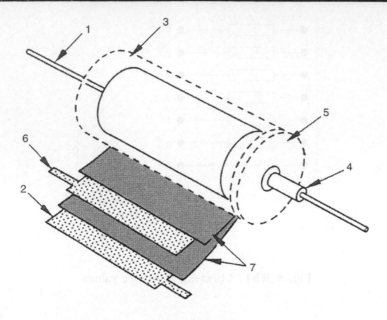

Fig. 8.5 Construction of a capacitor. 1. Negative connection; 2. Aluminium foil; 3. Metal case; 4. Positive connection; 5. Insulating disc; 6. Positive conducting strip; 7. Dielectric (insulation)

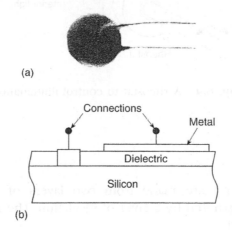

(a)

Connections

Metal

Dielectric

Silicon

(b)

Fig. 8.6(a) and (b) A ceramic and an I/C capacitor

Action of a capacitor

In the circuit shown in Figure 8.7(a), the battery voltage will be applied when the switch is closed. The capacitor voltage will not rise immediately to battery voltage because it takes a period of time owing to the movement of electrons. While the capacitor voltage is building up, current flows in the circuit until battery voltage is reached. This 'flow' of current only occurs when the voltage is changing.

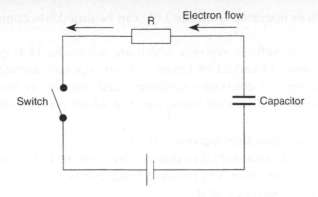

Fig. 8.7(a)   Charging a capacitor

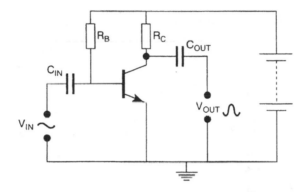

Fig. 8.7(b)   Coupling capacitors in a transistor circuit

When charged to battery voltage the capacitor will hold its electrical charge until it is provided with a circuit into which it can discharge. Storing an electrical charge is one use of a capacitor.

A coupling capacitor

If instead of applying d.c. to the capacitor an a.c. voltage is applied, the voltage is continuously changing. This means that an a.c. current will apparently flow through the capacitor. This apparent *current* is called displacement current and it may be thought of as a.c. current actually flowing around the complete a.c. circuit. Figure 8.7(b) shows a transistor amplifier that amplifies a small a.c. voltage into a larger one. The coupling capacitors separate the a.c. and d.c. parts of the transistor's operation.

TIMERS

The operation of a digital ECU is dependent on timing of the processes. This timing is provided by the 'processor clock' which may have a frequency measured in MHz. Any devices operated from the ECU can be timed, via counters, from this main clock.

Other processes in vehicle systems which are not under ECU control are also timed with precision. Electrical/electronic circuits operate according to precise physical laws (subject to constant conditions) and this means that timed operations can be produced in various ways, some of which are considered below.

Time constant of a capacitor-resistor circuit

As explained earlier, a capacitor takes time to charge up to the value of the voltage applied to it. The circuit shown in Figure 8.8 represents a practical circuit with a capacitor and some resistance in it.

When the switch is open the voltage across the capacitor will rise, as shown in the graph. The rate at which the capacitor charges is dependent on the size C (Farads) of the capacitor and the value of the resistance R (ohms).

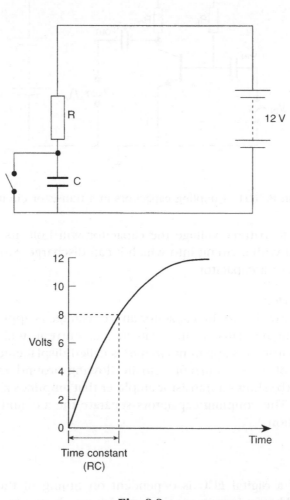

Fig. 8.8

The time it takes for the voltage to reach two-thirds of the applied voltage is known as the time constant T, and T = C × R.

As an example, take a 0.01 Farad capacitor and a 10 000 ohm resistor. The time constant T = 0.01 × 10 000, which gives 100 seconds.

If the applied voltage is 12 V it would take 100 seconds for the capacitor voltage to reach 8 V. It could also be said that when the voltage has reached 8 V the time from switching on is 100 seconds.

Such a CR (capacitive resistive) circuit can be constructed to provide a given time for operation of a device by the choice of suitable values of capacitance C (Farads) and resistance R (ohms).

Vehicle applications of timers include: courtesy light delay, heated rear window, variable periods of operation for front screen and rear wiper action, timed period for operation of heater plugs on diesel engines, to name just a few.

As stated above, the time taken for a capacitor to release its charge into a resistive circuit is dependent on the amount of resistance in the circuit. A simple circuit can be constructed to provide the delay that occurs in switching off interior lights. For example, the electronic circuit shown in Figure 8.9 operates the lights around the driver's door keyhole, ignition switch, and in the foot well to make it easier to get into the vehicle at night.

When the outside door handle, on the driver's side, is lifted the door handle switch is closed and the transistors Tr_1 and Tr_2 are switched on to light the entrance illumination. At the same time, the capacitor C_1 is charged up.

When the driver's side door handle is lowered the capacitor C_1 discharges into the circuit, keeping Tr_1 and Tr_2 operating, and hence keeping the entrance light on, until the capacitor has discharged to a level where the transistors cause the lamp current to stop flowing, and the entrance light goes out. In the case shown, which is for a Toyota model, the time is 5 seconds.

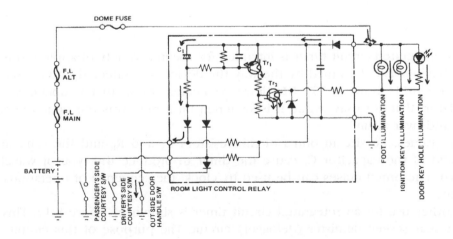

Fig. 8.9

Integrated circuit (I/C) timer

When a timer is constructed from individual capacitors and resistors it is said to be made from 'discretes', as opposed to being made in integrated circuit form.

A commonly used timer, known as the 555, is available as an I/C and some idea of its application will be gained from the following description.

Figure 8.10 shows the circuit that makes up the 555 timer. There are three resistors which make a voltage divider: two comparators, an RS flip-flop and a transistor.

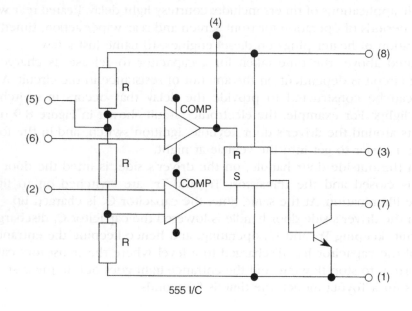

Fig. 8.10   An integrated circuit timer

The integrated circuit timer is intended for use in a variety of applications and its behaviour is determined by the way the device is connected into a circuit.

Figure 8.11 shows a 555 I/C timer connected up so that it operates as an oscillator; that is to say, while it is switched on it performs the same operation over and over again.

The values, i.e size in ohms of the resistors R_A and R_B and the capacitance (Farads) of the capacitor C, cause the timer to produce the type of waveform shown. The timed pulses can be used to 'clock' the sequence of operations in a circuit.

Another use for an integrated circuit timer is shown in Figure 8.12. This is a Rover rear screen demister (defogger) circuit. The purpose of this circuit is to avoid overloading the battery by preventing undue length of operation of the demister.

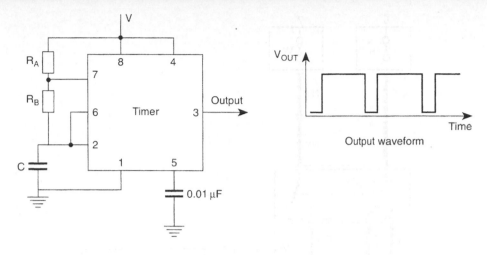

Fig. 8.11 A 555 I/C timer as an oscillator

The demist switch is used to energise the circuit. When pressure is removed from the switch, a spring opens the switch contacts to break the earth path of the timer circuit. The timer unit coil remains energised for approximately 10 minutes through the earth path on the black wire.

THE ZENER DIODE

Normally a diode only passes current when it is forward biased. If the polarity is reversed, i.e. it is reverse biased, the diode will not pass current. However, if the reverse bias voltage is sufficiently high the diode will be damaged.

The Zener diode is made in a special way so that it will pass current in the reverse direction. The voltage at which a Zener diode 'breaks down' to pass current in the reverse direction is known as the Zener voltage of the diode.

Zener diodes are available to suit a range of voltages and because each one has specific characteristics, e.g. the Zener voltage is precise, they are often used for voltage reference purposes.

Figures 8.13 (a) and (b) show the basic principle of the Zener diode. The circuit shown has a battery across which is a potential divider. There is also a circuit which contains a Zener diode and a lamp. In 8.13(a) the potential divider slider S is placed towards the negative end of the battery and the voltage applied to the Zener diode and lamp is too low to pass current through the Zener diode and the lamp does not light up.

In 8.13(b) the potential divider slider S has been moved towards the positive end of the battery. A higher voltage is now applied to the Zener diode. In this case it has reached the breakdown (Zener) voltage.

Battery current now flows through the Zener diode and lamp and back to the battery negative and the lamp is illuminated. In this application the Zener diode

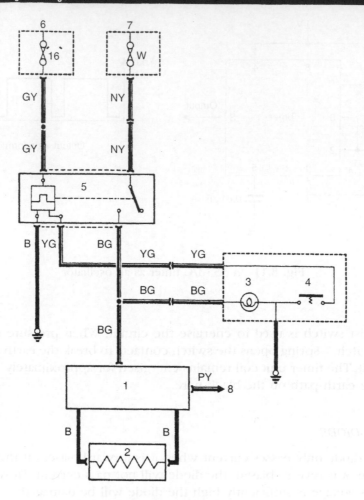

Heated rear window
1. Aerial amplifier
2. Heated rear screen
3. Warning light
4. Heated rear screen switch
5. Heated rear window timer
6. In car fusebox
7. Engine bay fusebox

Fig. 8.12 A timer in a heated rear window circuit (Rover)

is acting as a voltage sensitive switch. This Zener voltage is quite precise, which makes the Zener diode suitable for a variety of applications. Two such applications to motor vehicles are:

- a circuit protection device (load dump)
- a voltage reference in a voltage regulator

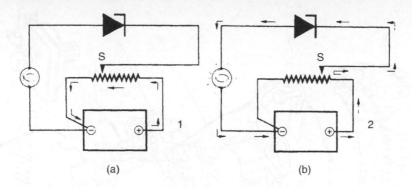

Fig. 8.13(a) and (b)   The Zener diode

Circuit protection

Vehicle circuits are subject to 'transient' voltages which arise from several sources. Those which interest us here are: load dump, alternator field decay voltage, switching of an inductive device (coil, relay, etc.) and over voltage arising from incorrect use of batteries when 'jump starting'.

Load dump occurs when an alternator becomes disconnected from the vehicle battery while the alternator is charging i.e. when the engine is running.

Figures 8.5 and 8.15 show a Zener diode as used for surge protection in an alternator circuit.

The breakdown (Zener) voltage of the diode is 10 to 15 volts above the normal system voltage. Such voltages can result if an open circuit occurs in the main alternator output lead when the engine is running. Other vehicle circuits, such as coil ignition, can also create inductive surges. Should such voltage surges occur

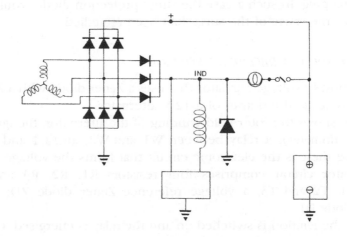

Fig. 8.14   Zener diode for circuit protection

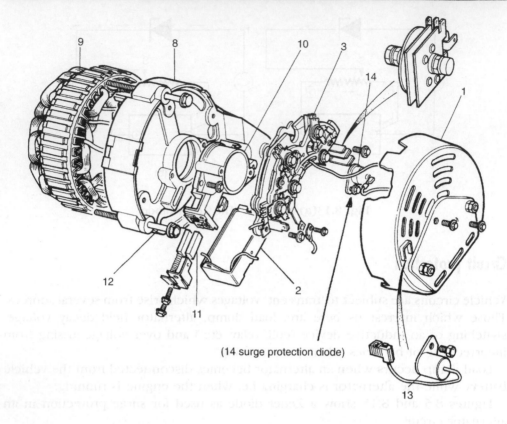

(14 surge protection diode)

Fig. 8.15 Voltage surge protection by Zener diode

they could damage the alternator circuits but, with the Zener diode connected as shown, the excess voltage is 'dumped' to earth via the Zener diode. Should there be such a voltage surge it may destroy the protection diode; the alternator would then cease charging. In such a case the surge protection diode would need to be replaced, after the cause of the surge had been remedied.

THE ZENER DIODE AS A REFERENCE VOLTAGE

Figure 8.16 shows a voltage regulator that uses a Zener diode with a Zener voltage of 14.2 V to regulate the output of a 12 V alternator.

This circuit shows the rotor field winding of the alternator, the ignition switch (next to the ammeter), a relay between W1 and W2, and C1 and C2, and the battery. Above these is the electronic circuit that forms the voltage regulator.

The regulator circuit comprises four resistors R1, R2, R3 and R4; three transistors T1, T2 and T3; a voltage reference Zener diode ZD; and a surge protection diode D1.

As shown, the ignition is switched on and the relay is energised. Current flows from the battery positive, through the relay contacts and resistors R1 and R2, and back to the battery via the earth return.

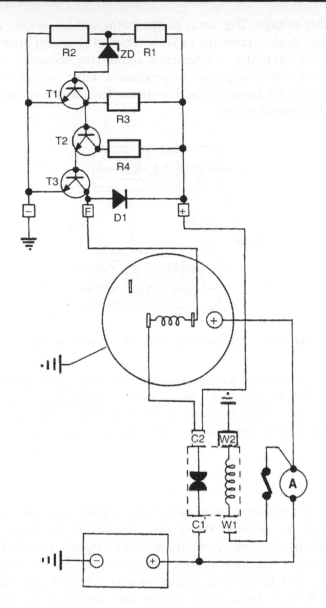

Fig. 8.16 The Zener diode in a voltage regulator diode circuit

A proportion of current also flows through resistor R3 to the base terminal of transistor T2. This 'switches' T2 on and permits current to flow through the collector-emitter circuit of T2 via R4. This, in turn, 'switches on' T3 and permits current to flow through the rotor field. This action permits the alternator output to rise and, when the battery terminal voltage reaches 14.2 V, the Zener diode conducts current to the base of T1. The application of voltage to the base of T1 switches it on and current flows through R3 and the collector-emitter of T1 to earth and T2 and T3 are switched off. This results in a reduced field current and a drop in alternator voltage. This drop in alternator voltage stops current flowing through the Zener diode. Transistor T1 is switched off, current now flows through T2 and T3 and the alternator voltage rises again. This sequence of operation is repeated at the necessary frequency to produce a stable voltage supply from the alternator. The diode D1 protects the transistors from surge voltage which arises when transistor T3 switches off.

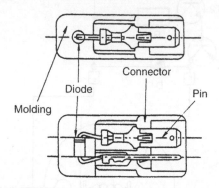

Fig. 8.17 A protection diode in a cable connector

Figure 8.17 shows another form of circuit protection where a diode is built into a cable connector. This reduces the risk of damage from reversed connections and it is evident that one should be aware of such uses because a continuity test on such a connector will require correct polarity at the meter leads.

Drivers

In the section of Chapter 6 that deals with inputs and outputs at the ECU reference is made to drivers. These drivers convert the logic levels of the controller output to the higher electrical power signals required by actuators, such as fuel injectors, stepper motors, ABS actuators, etc. As these drivers will generally be required to provide higher electric current and voltage we may expect to be looking at devices that amplify and are robust in construction.

It is considered appropriate for vehicle repair purposes to know about three types of drivers: namely, Darlington, display and stepper motor, so we will now consider some details of each of these.

DARLINGTON TRANSISTOR (DARLINGTON PAIR)

In earlier sections we have seen that the basic transistor has a very small base current which permits a much larger current to flow in the collector-emitter circuit. For a specific transistor the base current may be 0.1 mA and the collector current 20 mA. The collector current divided by the base current gives a figure of 200. This is known as the current gain of the transistor.

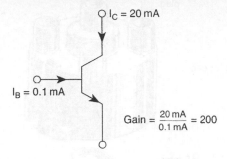

Fig. 8.18 Current gain – a simple transistor

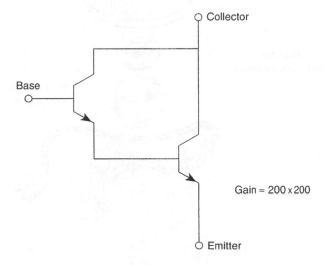

Fig. 8.19 The Darlington pair (a driver)

In the Darlington transistor two transistors are coupled together (cascaded) so that the emitter current of the first transistor becomes the base current of the second one. In this way the current gain is the first transistor's gain multiplied by the second transistor's gain.

An electronic ignition system

TRANSISTORS APPLIED TO AN IGNITION SYSTEM

Figure 8.20 shows an electro-magnetic pulse generator (variable reluctance) of the type used in electronic ignition systems.

This pulse generator produces a waveform of approximately the same shape as the one shown in Figure 8.21. This waveform has to be processed to make sure it is suitable for operating the ignition circuit.

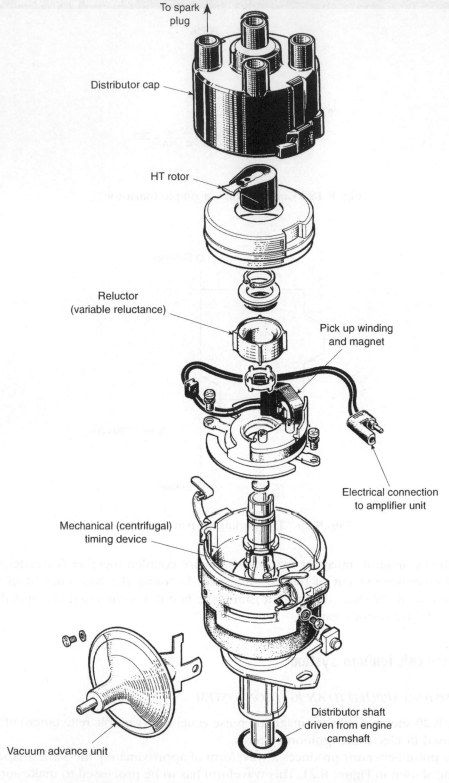

To spark plug

Distributor cap

HT rotor

Reluctor (variable reluctance)

Pick up winding and magnet

Electrical connection to amplifier unit

Mechanical (centrifugal) timing device

Vacuum advance unit

Distributor shaft driven from engine camshaft

Fig. 8.20 An ignition distributor with a pulse generator

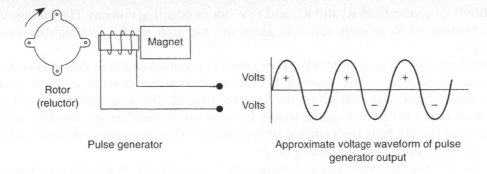

Rotor
(reluctor)

Magnet

Volts

Pulse generator

Approximate voltage waveform of pulse
generator output

Fig. 8.21 The generator voltage waveform

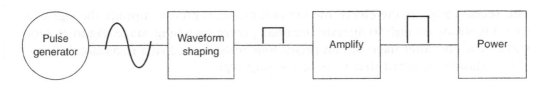

Fig. 8.22 The processing stages of the ignition circuit

For precise switching 'on and off' of current in the coil primary circuit it is necessary to convert this waveform to a more suitable rectangular form.

The processing stages that the pulse generator voltage passes through in order to generate the high voltage spark at the spark plug are shown in Figure 8.22.

The conversion of the curved wave from the pulse generator into the rectangular form required for accurate switching can be performed by a circuit known as a Schmitt trigger. In the simplified circuit shown in Figure 8.23 the Schmitt trigger comprises two bi-polar transistors (T_1 and T_2) and five resistors.

When the pulse generator voltage is positive (above a certain value), T_1 will conduct through R_1, the collector-emitter and R_2 to earth. This has the effect of

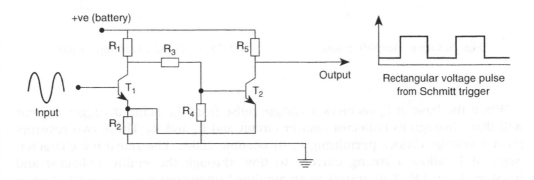

+ve (battery)

R_1 R_3 R_5

T_1 T_2

Input R_2 R_4

Output

Rectangular voltage pulse
from Schmitt trigger

Fig. 8.23 A Schmitt trigger circuit

removing current from R_3 and R_4, and T_2 is not conducting current. The voltage at the bottom of R_5 is high and this gives the top part of the rectangular wave pattern.

When the pulse generator wave is negative T_1 is not conducting; this means that voltage is applied to the base of T_2. The voltage at the bottom of R_5 is now practically zero, and this gives the bottom part of the rectangular wave. The vertical edges of the waveform would be seen on an oscilloscope because they represent the fall from high voltage to low voltage. They are sometimes referred to as the rising edge and the falling edge.

The rectangular wave pattern shown here is idealised. Practical waveforms, from the type of circuit shown, tend to be a little less sharply defined.

AMPLIFICATION

The second stage of the electronic ignition circuit is used to amplify the signal so that it is strong enough to operate the final, power switching, stage. Figure 8.24(a) shows the amplifier stage; it comprises four resistors and two transistors (T_3 and T_4). It should be noted that T4 is of the pnp type.

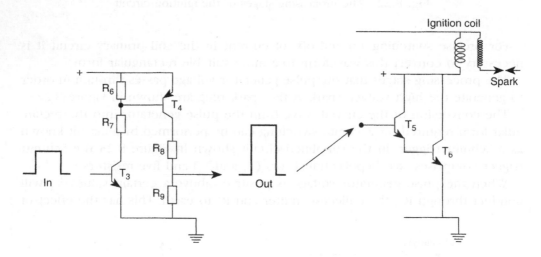

Fig. 8.24(a) Amplifier stage   **(b)** Power switch, Darlington pair

When the base of T_3 receives a voltage pulse from the Schmitt trigger current will flow through its collector-emitter circuit and R_6 and R_7. These two resistors form a voltage divider permitting T_4 to become active. The transistor characteristics of T_4 allow a strong current to flow through the emitter-collector and resistors R_8 and R_9. This results in an amplified (increased power) signal of good switching shape which is used to operate the power switching stage.

THE POWER STAGE

The power switching stage consists of the two transistors, T_5 and T_6. The amplified signal is applied to the base of T_5: this allows a large current to flow through the T_5 collector-emitter to the base of T_6. The current gain of T_6 (the coil primary circuit current) is then passed through the collector-emitter of T_6 to earth. The power switching stage is shown in Figure 8.24(b) and it is the Darlington pair. These 'electronic stages' are normally encapsulated in a module which is repaired by replacement.

As with all other vehicle electronic systems, it is possible to conduct tests that enable one to be reasonably certain that the pulse generator and ignition coil are working properly so that, by elimination, an electronic module defect can be detected. Some of the practical tests that can be performed are now shown.

Figure 8.25 shows an ohm-meter connected to the pulse generator terminals. If the test shows infinity on the ohms scale, a broken coil is indicated, and if the reading is very low a shorted coil is indicated. In either case the pulse generator is defective.

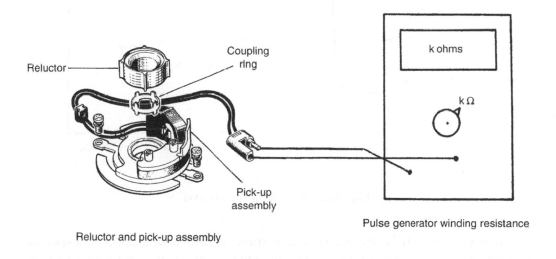

Reluctor and pick-up assembly

Pulse generator winding resistance

Fig. 8.25 Testing the pulse generator coil

The ignition coil can be checked as shown in Figure 8.26. The secondary winding resistance should be high, probably 10 k ohms. (*Note.* In some ignition coils the secondary winding is earthed. In such cases the check for secondary winding resistance would be made between the HT connection and the coil earth.)

The primary winding resistance should be very low (the meter needs to be set to a very low ohms scale), 0.5 ohms to 1 ohm.

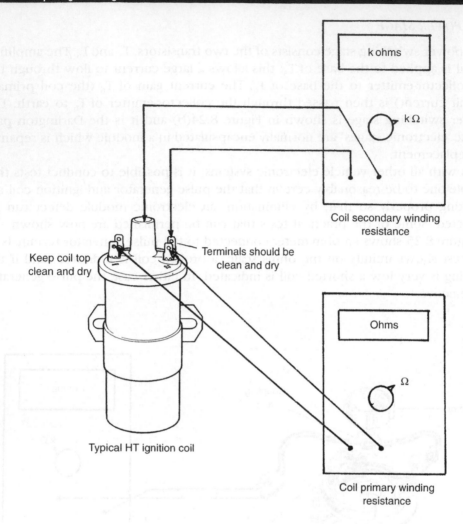

k ohms

kΩ

Coil secondary winding
resistance

Keep coil top
clean and dry

Terminals should be
clean and dry

Ohms

Ω

Typical HT ignition coil

Coil primary winding
resistance

Fig. 8.26 Checking the HT coil

The tests shown here are of a general nature, i.e. they do not apply to a specific
system and, as with most other work on vehicle systems, it is important to have
the figures that relate to the specific system being worked on.

However, the tests do show the practical nature of tests that can be applied
with the aid of a good quality multi-meter, sound knowledge of circuits and the
relevant information.

A WORD OF WARNING

Ignition systems generate many thousands of volts at the HT side. Great care must
be exercised in order to avoid electric shock. In some cases such electric shock
can be fatal, and in other cases it may cause a sharp nervous reaction which, in
turn, can lead to an accident.

From previous page
B

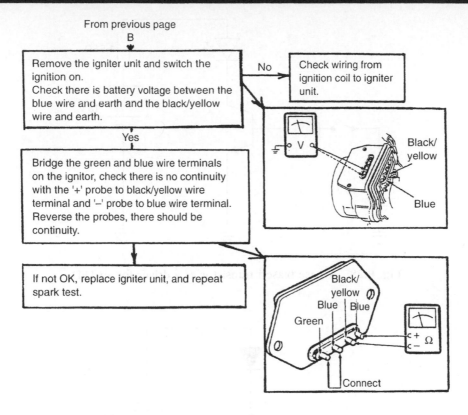

Remove the igniter unit and switch the ignition on.
Check there is battery voltage between the blue wire and earth and the black/yellow wire and earth.

No → Check wiring from ignition coil to igniter unit.

Yes

Bridge the green and blue wire terminals on the ignitor, check there is no continuity with the '+' probe to black/yellow wire terminal and '−' probe to blue wire terminal. Reverse the probes, there should be continuity.

Black/yellow

Blue

If not OK, replace igniter unit, and repeat spark test.

Black/yellow
Blue Blue
Green

+ Ω
−

Connect

Fig. 8.27   Testing an igniter (electronic module)

Testing the electronic module (Figure 8.27)

This extract from a fault tracing chart for a Rover 213 ignition system shows tests that can be used on the 'igniter' unit which is what the electronic module is sometimes called. Other systems will be different, but this example serves to show that when the service information is available the 'electronic' system checks are well within the compass of ability of a good vehicle technician.

An LED driver circuit

In this section two circuits that are used as LED drivers are considered: these are the base biased transistor and the emitter biased transistor.

THE BASE BIASED LED DRIVER

As the circuit is shown in Figure 8.28 the switch is open and there is no base current. When the switch is closed the transistor goes into hard saturation, current flows through the LED and light is given out. In this way the LED is driven by the transistor. An alternative is to use an emitter biased LED driver.

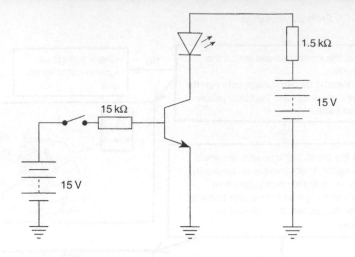

Fig. 8.28 A base biased transistor as a driver for an LED

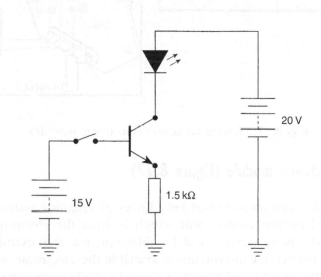

Fig. 8.29 An emitter biased transistor used as a driver for an LED

Figure 8.29 shows an alternative arrangement for an LED driver. Here the resistor is on the emitter side of the transistor.

STEPPER MOTOR DRIVER CIRCUIT

The stepper motor driver circuit is shown in Chapter 2.

A TRANSISTOR AS A VOLTAGE AMPLIFIER

Circuit elements such as resistors are known as passive devices because they do not produce, gain or store energy. As we have seen, in the Darlington pair for example, transistors produce current gain, and because they act in this way transistors are known as active devices.

In addition to its use as a switch and as a device to produce current gain the transistor can also be built into a circuit to produce voltage gain. These properties are useful in vehicle systems because of the range of inputs from various sensors, and the outputs which the ECU produces to initiate actions from actuators such as injector pulses, stepper motor movement, ABS modulator valve motion, etc.

The circuit shown in Figure 8.30 is a transistor circuit that produces voltage gain. The d.c. voltage value and the resistor values are chosen (biased) so that the transistor operates as required. The application of an a.c. voltage V_{in} at the input terminals of the circuit will produce a larger voltage V_{out} at the output terminals. The ratio V_{out}/V_{in} is known as the voltage gain of the amplifier. When viewed on an oscilloscope the output waveform looks just like a magnified version of the a.c. input waveform.

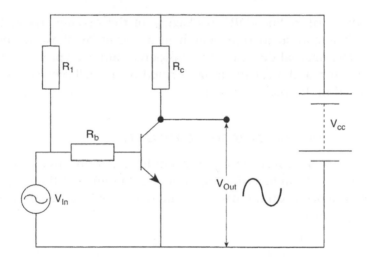

Fig. 8.30 A voltage amplifier circuit

The operational amplifier

The principle of voltage amplification as shown in Figure 8.30 is used in an integrated circuit known as an operational amplifier (op-amp). Figure 8.31 shows the packaged article with its pins which make it readily available to build into circuits.

The op-amp gives a very high open circuit voltage gain. By the use of external circuits, feed back resistors, etc, and by using the inverting and non-inverting

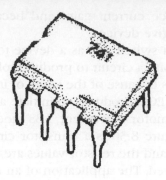

Fig. 8.31 The 741 op-amp

inputs differently, the op-amp can be made to perform a wide range of electronic functions.

Although the dual in line (DIL) packaging of the op-amp has eight external connections, it is normal to represent it in a circuit by showing just the four connections. The internal detail is not considered here except to say that several transistors, resistors and a capacitor are formed on a small piece of silicon (chip) which is housed in the plastic casing.

CIRCUIT SYMBOL FOR THE OPERATIONAL AMPLIFIER

A signal applied to the inverting input is amplified and inverted at the output. A signal applied to the non-inverting input is also amplified at the output, but it is not inverted. It depends on the nature of the output voltage required as to which input terminal is used.

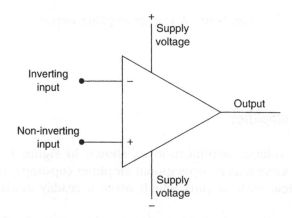

Fig. 8.32 The circuit symbol for an operational amplifier

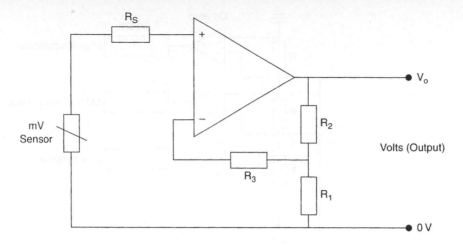

Fig. 8.33 An op-amp in a temperature sensor circuit

USES OF THE OP-AMP

Using the op-amp as a d.c. amplifier

A small, slowly varying voltage from a sensor can be amplified to a suitable form by the use of an op-amp that has been connected into the circuit to act as a d.c. amplifier, as shown in Figure 8.33.

COMPARATOR

When an op-amp is connected as shown in Figure 8.34(a) it can be used as a comparator. When the input voltages are unequal the output of the op-amp is low

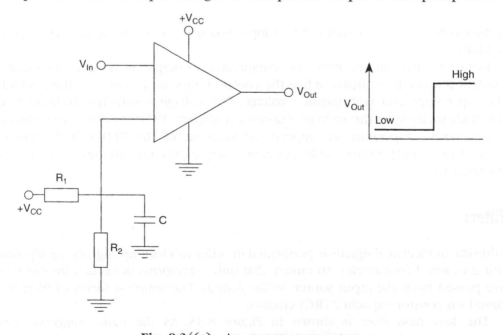

Fig. 8.34(a) An op-amp comparator

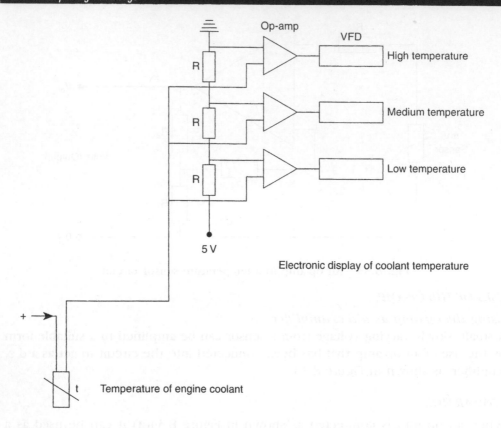

Fig. 8.34(b)   Comparators in a VFD display

(effectively zero), and when the two input voltages are equal the op-amp output is high.

Figure 8.34(b) shows how the comparator principle is used in an engine coolant temperature display. When the coolant temperature sensor voltage is high the top comparator will output a voltage. This voltage is sufficient to illuminate the high segment of the vacuum fluorescent display (VFD). As the sensor output varies with temperature an appropriate segment of the VFD will illuminate. Several such comparators can be made on a single I/C chip and they are relatively inexpensive.

Filters

Filtering of electrical signals is performed in order to 'clean up' signals, by filtering out unwanted frequencies, to ensure that only waveforms of certain frequencies are passed from the input source to the output. The simplest forms of filter are based on resistor capacitor (RC) circuits.

The low pass filter is shown in Figure 8.35. As the name suggests, low frequency signals will be passed. This is because the capacitor is like an open

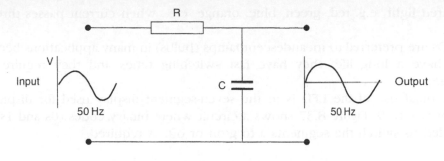

Fig. 8.35 The low pass filter (lo-pass)

switch at low frequency and a closed switch at higher frequency. Examination of the circuit will show that low frequency signals will pass from input to output, through the resistor. At higher frequency the capacitor conducts a.c. and effectively short circuits the supply thus preventing it from reaching the output.

The other form of simple filter is the high pass filter. This again relies on resistance and capacitance, but in this circuit the resistor and capacitor are in different positions: here the lower frequencies will be 'blocked' by the capacitor. As the frequency rises, a point will be reached at which the capacitor starts to conduct a.c. The input and output are thus connected to 'pass' higher frequencies. The actual frequency at which the filter will pass current is dependent on the values of R and C.

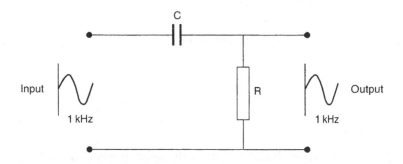

Fig. 8.36 The high pass filter (hi-pass)

Light emitting diodes

The light emitting diode (LED) is a device which gives out light when current passes through it. Ordinary diodes made of silicon do not give out light because silicon is an opaque material. When diodes are made from other elements, such as arsenic, phosphorous and gallium, LEDs are produced which give out different coloured light, e.g. red, green, blue, orange, etc., when current passes through them.

LEDs are preferred to incandescent lamps (bulbs) in many applications because they have a long life, they have fast switching times and they require low voltage.

A typical use of the LED is in the seven-segment display used for displaying numbers 1 to 9. Figure 8.37 shows a circuit where binary codes (0s and 1s) are decoded to switch the segments a to g on or off, as required.

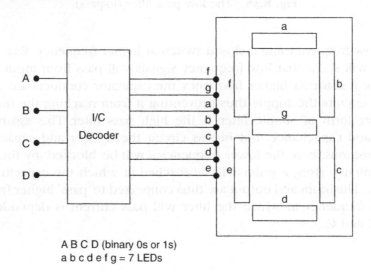

A B C D (binary 0s or 1s)
a b c d e f g = 7 LEDs

Fig. 8.37

THE PHOTO DIODE

When a p–n junction is bombarded with light energy it dislodges valence electrons and this gives rise to a reverse current through the p–n junction. The greater the intensity of light that falls on the p–n junction the greater is the reverse current produced. This effect means the photo diode can be used as a sensor which produces a signal current that is related to the amount of light energy falling on it.

Photo diodes are constructed with a window that permits light to pass through the package to the p–n junction so that the effect of the light on the junction is optimised. The amount of reverse current produced is small, but in a carefully designed circuit the effect is such that the output of a sensor can be made accurately to represent the intensity of light falling on the sensor. The approximate effect is shown in Figure 8.38.

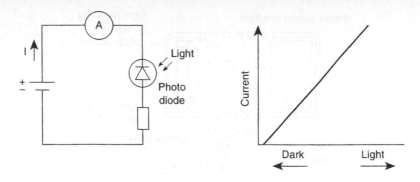

Fig. 8.38

Automatic control of vehicle lights

An interesting application of the photo diode sensor is to be found in the Toyota system shown in Figure 8.39.

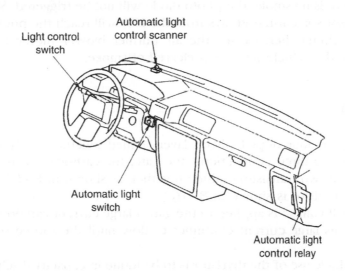

Fig. 8.39 Photodiode as sensor in automatic light system (Toyota)

Figure 8.39 shows a left-hand drive vehicle. The light sensor is placed in a position where it will best detect the ambient light level at any given time so that driving lights can be switched on automatically.

Air purifier system

Some vehicles in the Toyota range are equippped with air purifier systems. Part of this air purification system is a smoke detector. In this particular application the smoke detecting sensor utilises a photo diode and an LED.

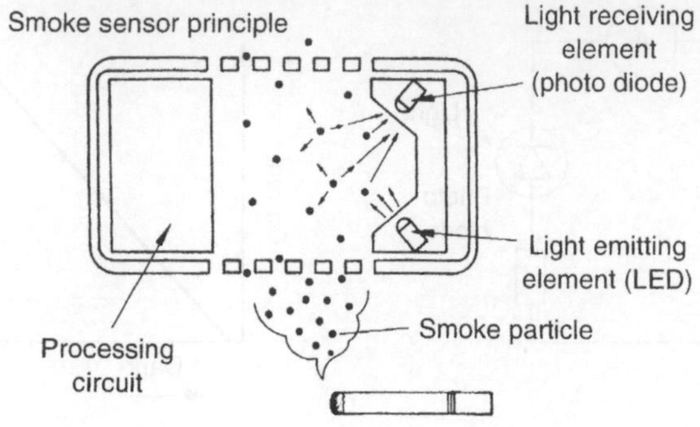

Smoke sensor principle

Light receiving element (photo diode)

Light emitting element (LED)

Smoke particle

Processing circuit

Fig. 8.40   Principal of smoke detector sensor (Toyota)

The sensor is placed in a suitable position in the vehicle so that smoke, such as cigarette smoke, will freely enter through the slots in the sensor casing. The LED part of the sensor sends out infrared rays at intervals when supplied with a pulsed current. If there is no smoke the photo diode will not be triggered. Should smoke enter the sensor some infrared rays from the LED will reach the photo diode. The photo diode circuit then causes the air purifier blower motor to commence operation and the vehicle interior is cleared of smoke.

The thyristor

The thyristor has four doped silicon layers joined together as shown in Figure 8.41(a). There are three connections: the gate, the cathode and the anode. The device acts like two transistors joined together, as shown in 8.41(b). The circuit symbol for a thyristor is shown in 8.41(c).

When a small voltage is applied to the gate a large current can flow from anode to cathode. This large current continues to flow until the source supplying it is switched off.

A motor vehicle use of the thyristor is to be found in capacity discharge ignition systems. The principle of such a system is shown in Figure 8.42.

The electronic module produces a voltage source of approximately 400 V; this 400 V is used to charge the capacitor. When the engine's ignition timing circuit is required to produce a spark, at the spark plug, the thyristor gate is energised; the thyristor then permits the capacitor to discharge through the coil primary circuit and, by mutual induction, a high voltage (approximately 40 kV) is produced in the coil secondary winding. It is a very high voltage system and because of the danger of electric shock great care must be taken when working on such systems.

Another use of the thyristor, or silicon controlled rectifier (SCR), is to protect expensive integrated circuits from overvoltage. Such excess voltage can arise from a power supply and the thyristor is placed across the power supply output to short-circuit the output and protect the I/C which is being supplied.

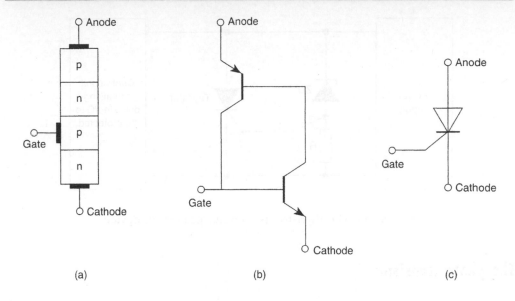

Fig. 8.41 The thyristor

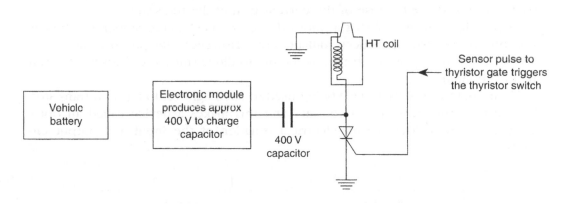

Fig. 8.42 A capacity discharge ignition system

When the thyristor is used in the way shown in Figure 8.43 it is sometimes known as a 'crowbar' because it is equivalent to placing a metal conductor across the power supply output leads.

In emergency situations the thyristor is triggered very quickly (1μs in some cases) and it thus provides an effective protection for the integrated circuit. When a power supply is fitted with the crowbar device it is necessary to build a fuse or some other current limiting device into the power unit to protect it from excess current during emergency operation when the thyristor is conducting.

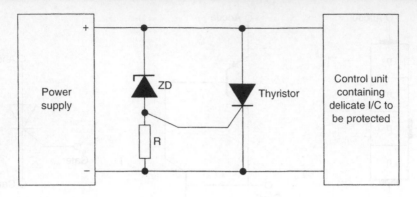

Fig. 8.43   The thyristor as a circuit protection device

The photo transistor

In the description of the photo diode it was stated that light energy falling on the p–n junction produces a reverse current. A similar effect occurs in the photo transistor. In a transistor with an open base a small current, arising from thermal effects, exists in the collector circuit. When the base-collector junction is exposed to light additional current is produced. This current is greater than that produced by the photo diode because of the current gain of the transistor.

In practice, this means that a photo diode produces micro amperes, whereas the photo transistor produces milliamperes. However, the photo transistor is slower in switching on or off than is the photo diode (micro seconds rather than nano seconds).

Figure 8.44(a) shows an LED being used to shed light on to a photo transistor. This causes the transistor to conduct and this makes a device called an opto-coupler. Such devices permit electrical isolation of the input and output circuits.

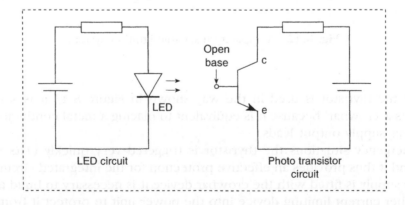

Fig. 8.44(a)   A photo-transistor in an opto-coupler

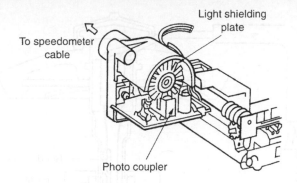

Fig. 8.44(b)   Speed sensor (Toyota)

Figure 8.44(b) shows an application of a photo transistor as part of a speed sensor. The sensor is essentially an opto-coupler.

The light shielding plate is rotated by the speedometer cable. This interrupts the light source and the transistor side of the opto-coupler produces a current supply which is switched on and off at a frequency related to vehicle speed.

Sensors

THE REED SWITCH

A reed switch consists of a set of contacts inside an evacuated (or filled with inert gas) tube. The movable contact is attracted into contact with the mating contact by means of a magnetic field. When the field is reversed the switch contacts separate. In this way the switch can operate at about 500 cycles per second and have a long life. Figure 8.45(a) shows the type of symbol used in circuit diagrams to denote a reed switch; and Figures 8.45(b) and 8.45(c) show vehicle applications of the device.

Fig. 8.45(a)   Reed switch symbol

FLUID LEVEL

Figure 8.46(b) shows how the reed switch can be incorporated into a warning light system to indicate when the brake fluid in the reservoir has reached a critical level. It is evident that there are other fluid levels that could be indicated by the same means.

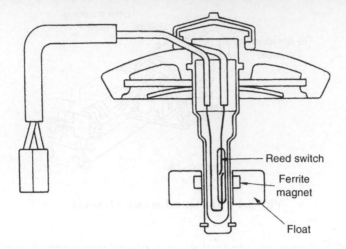

Fig. 8.45(b) Reed switch fluid level sensor

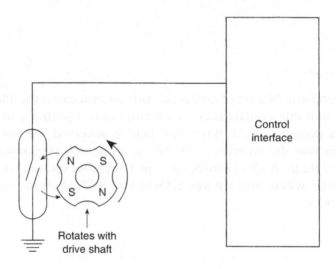

Fig. 8.45(c) A reed switch speed sensor

SPEED SENSOR

Figure 8.45(c) shows how the reed switch can be used to provide an indication of speed of rotation of a shaft.

The magnet rotates with the shaft and as the different magnetic poles pass the critical part of the reed switch an 'on-off' cycle is completed. The frequency at which this switching occurs can then be used to compute an equivalent road speed.

Manifold absolute pressure (MAP)

The pressure of the air, or mixture, in the engine intake manifold varies with the load on the engine. For example, with the throttle valve fully closed and the engine being used for downhill braking, the manifold pressure will be very low (near perfect vacuum); with the throttle fully open and the vehicle accelerating up an incline the manifold pressure will be higher, i.e very little vacuum. The pressure inside the intake manifold is known as absolute pressure (i.e. it is measured from absolute zero pressure, or complete vacuum, upwards). Manifold absolute pressure gives a very good guide to (analogy of) engine load and its measurement plays an important part in engine control systems.

Additionally, an accurate guide to the amount of air entering an engine can be computed from volume, manifold absolute pressure and temperature. Systems which make use of this method are known as speed density systems.

The accurate measurement of manifold pressure plays an important part in engine control strategies and the MAP sensor is the device that does this.

MAP sensors are available in several different forms. For our purposes it will suffice to examine one of these to give an insight into the method of operation, and the types of tests that can usefully be performed on them.

MAP SENSOR

The manifold absolute pressure (MAP) sensor shown in Figure 8.46 receives a 5 V supply from the ECU. Variations in manifold pressure (vacuum) cause the small silicon diaphragm to deflect. This deflection alters the resistance of the resistors in the sensor's bridge circuit and the resulting electrical output from the bridge circuit is proportional to manifold pressure.

By the use of an electronic amplifier the voltage is increased so that the sensor output lies between 0 V and 3 V, according to pressure, as shown on the graph.

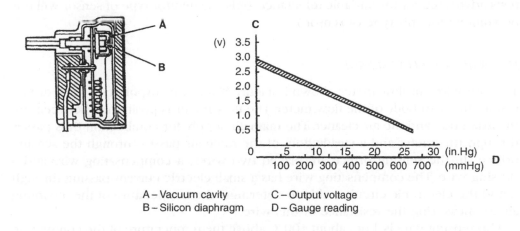

A – Vacuum cavity   C – Output voltage
B – Silicon diaphragm   D – Gauge reading

Fig. 8.46 A manifold absolute pressure sensor

(Note the double line: this indicates that a considerable variation in voltage, for a given pressure (vacuum), may be expected.)

It is evident that a number of checks may be performed on such a sensor. For example, if the correct input voltage is known to be 5 V (±0.35 V), a simple voltage check at the sensor input terminal will verify this. In a similar way, a voltage check at the sensor output terminal, together with a vacuum gauge test on the engine, will permit an assessment of the sensor's performance. Such checks are readily performed by the use of the standard type of oscilloscope based analysers.

Once again I must urge caution because the actual information relating to a particular vehicle must be to hand prior to attempting such tests.

An important point to draw out of this chapter is that sensors often produce quite small amounts of electrical energy. If the circuit between the sensor and the ECU is not perfect some of this energy will be lost and this affects the performance of the whole system.

HALL EFFECT SENSOR

Figure 8.47 shows a Hall type sensor that is used in an electronic ignition system. It relies for its operation on the 'Hall' effect, named after the man who first enunciated the principle.

In the application shown, the Hall element, in which an emf (voltage) is generated, is placed opposite to the magnet. When the iron of the rotating vane is between the magnet and the Hall element the magnetic field is strong and a voltage is produced by the sensor. As the vane rotates out of the gap the magnetic flux falls and the voltage generated also falls. This happens at the same frequency as that at which iron parts of the vane pass through the sensor gap. In the case shown, that is four times per revolution of the shaft.

The voltage produced is small, of the order of millivolts, and requires amplification to make it suitable for use in a vehicle system. (It should be evident that the tests advocated for the variable reluctance 'pulse' generator type of sensor will not be suitable for this type of sensor.)

HOT WIRE AIR FLOW SENSOR

The 'hot wire' air flow meter, shown in Figure 8.48(a), incorporates a small orifice inside the main body of the flow meter. The flow meter is positioned between the throttle body and the air cleaner. The main air supply for combustion thus passes through this meter and a steady flow of the same air passes through the sensing orifice. In the sensing orifice are placed two wires, a compensating wire and a sensing wire. The compensating wire has a small electric current passing through it and the electronic circuit is able to determine the temperature of the incoming air by measuring the resistance of this wire.

The sensing wire is 'hot', about 100°C above the temperature of the compensating wire. This temperature is maintained by varying the current flowing through

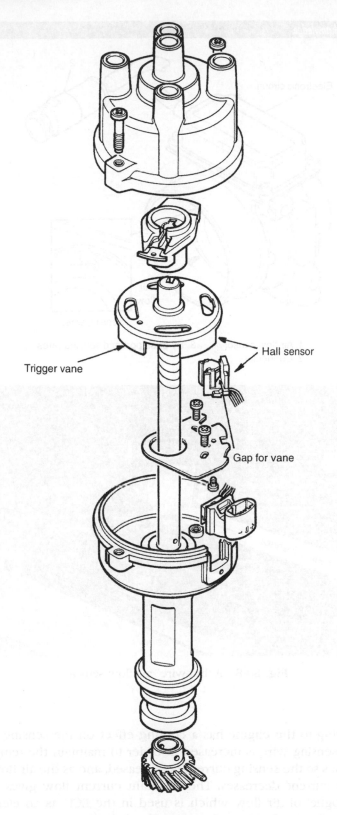

Trigger vane

Hall sensor

Gap for vane

Fig. 8.47 A Hall type sensor

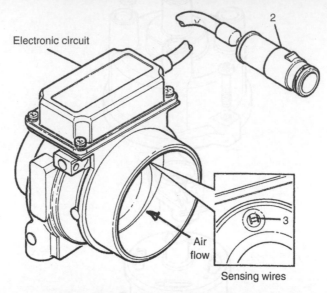

Electronic circuit

Air
flow

Sensing wires

1. Air flow sensor   3. By-pass port and sensing wires
2. Connector

(a)

Air Flow Sensor

(b)

Fig. 8.48 A 'hot-wire' air flow sensor

it. The air flowing to the engine has a cooling effect on the sensing wire so the current in the sensing wire is increased in order to maintain the temperature. As air flow increases so the sensing current is increased, and as the air flow decreases so the sensing current decreases. The resultant current flow gives an accurate electrical 'analogue' of air flow which is used in the ECU as an element of the controlling data for the fuel injection system.

The air-flow meter is normally placed between the air filter and the throttle body, as shown in Figure 8.48(b).

An important element of any test strategy for fault finding is testing the integrity of interconnecting circuits. This short review of a few vehicle type sensors shows that in general they produce a small electrical energy output. If any of this energy is lost in the circuit between the sensor and the ECU it may have a serious effect on the operation of the whole system.

9
Digital electronics and logic devices

In Chapter 2 the outline diagram of an ECU showed the engine speed sensor signal, the oxygen sensor signal, etc., connected to an interface and an 'analogue to digital' (A/D) converter. This conversion is required because the computing part of the ECU operates on 0s and 1s that are passed around the data buses.

When vehicle ECUs are networked, as described in Chapter 7, the data that passes between the ECUs does so in digital form. These two features provide the justification for including an introduction to some digital and computer 'logic' devices.

In a digital system the signals (voltages) can have only certain values, e.g. 0 or 1 (off or on), whereas in analogue systems the signals (voltages) may have any value within a specified range. For example, a digital system may be designed to have two possible output values, 0 V (off) and 5 V (on).

An analogue system, for example such as a 12 V charging system, may have an output of between 0 V and 15 V, and the actual voltage for a reading of 10 V may be approximately 10, depending on the accuracy of the measuring instrument.

Because digital systems work with precise values they can be used for purposes such as multiplying two 5-digit numbers. The resulting 10-digit product will be precise.

If an analogue multiplier is used for the same purpose, the result may contain a small percentage error which could be of critical importance in a computing process.

Using transistors to make logic devices

Transistors possess properties which make them suitable for use in building up 'logic' devices. Before examining some logic devices, it is necessary to look at switching properties of transistors.

SWITCHING TIMES OF TRANSISTORS

As mentioned earlier, in the description of the Schmitt trigger, the converted waveform is not exactly rectangular. A partial explanation of why this should be so is given in the following introduction to switching times of transistors.

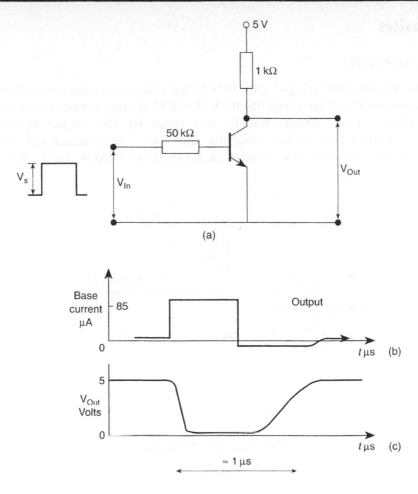

Fig. 9.1(a), (b) and (c)

Figure 9.1(a) shows a bipolar transistor switch. The switching voltage pulse V_s has the square waveform shown. When the voltage input V_{in} goes high, to V_s, base current immediately starts to flow into the transistor. This causes the base voltage to rise and when it reaches about 0.7 V, collector current starts to flow. The base voltage does not rise instantaneously because of capacitive effects at the collector-base junction and the emitter-base junction; 9.1(b) shows the approximate behaviour of base current and 9.1(c) shows the resultant shape of the waveform of the output voltage V_{out}. These give rise to 'turn on times' and 'turn off times' for transistors, which are very small, e.g. measured in nano seconds. However, these small delay times are the reason why the 'square' wave has slightly 'rounded' corners.

The switching properties of transistors make them ideal for use in circuits such as control units which use computer logic.

Logic devices

THE RTL NOR GATE

Figure 9.2 shows how a 'logic' gate is built up from an arrangement of resistors and a transistor. There are three inputs A, B and C. If one or more of these inputs is high (logic 1), the output will be low (logic 0). The output is shown as A + B + C with a line, or bar, over the top; the + sign means OR. Thus the A + B + C with the line above means 'not A or B or C' (NOR: NOT OR).

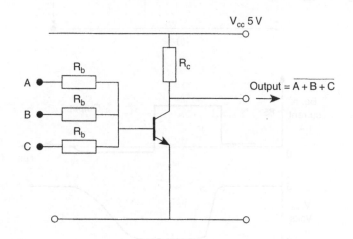

Fig. 9.2 RTL NOR gate

The base resistors R_b have a value that ensures that the base current, even when only one input is high (logic 1), will drive the transistor into saturation to make the output low (logic 0). [RTL stands for resistor transistor logic.]

TRUTH TABLES

Logic circuits operate on the basis of Boolean logic, and terms like NOT, NOR, NAND, etc. derive from Boolean algebra. This need not concern us here, but it is necessary to know that the input–output behaviour of logic devices is expressed in the form of a 'truth table'. The truth table for the NOR gate is given in Figure 9.3.

In computing and control systems, a system known as TTL (transistor to transistor logic) is used. In TTL logic 0 is a voltage between 0 and 0.8 V. Logic 1 is a voltage between 2.0 V and 5.0 V.

In the NOR truth table, when the inputs A and B are both 0 the gate output, C, is 1. The other three input combinations each give an output C = 1.

A range of other commonly used logic gates and their truth tables is given in Figure 9.4.

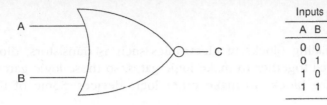

Inputs		Outputs
A	B	C
0	0	1
0	1	0
1	0	0
1	1	0

Fig. 9.3   NOR gate symbol and truth table

Type of logic gate	USA symbol	UK symbol	Truth table
AND			Inputs / Outputs A B / X 0 0 / 0 0 1 / 0 1 0 / 0 1 1 / 1
OR			Inputs / Outputs A B / X 0 0 / 0 0 1 / 1 1 0 / 1 1 1 / 1
(NOT AND) NAND			Inputs / Outputs A B / X 0 0 / 1 0 1 / 1 1 0 / 1 1 1 / 0
NOT inverter			Inputs / Outputs A / X 0 / 1 1 / 0
(NOT OR) NOR			Inputs / Outputs A B / X 0 0 / 1 0 1 / 0 1 0 / 0 1 1 / 0

Fig. 9.4   A table of logic gates and symbols

Other logic devices

Just as the basic building blocks of electronics such as transistors, diodes and resistors can be joined together to make logic gates, so those logic gates can be used as basic building blocks to make other logic devices. Some of these are introduced below.

FEEDBACK

The basic inverter shown in Fig. 9.5(a) produces an output of 1 when the input is 0 and an output of 0 when the input is 1.

When the output is fed back to the input as shown in Fig. 9.5(b) the output oscillates between logic 0 and logic 1 as shown. The frequency of this oscillation is dependent on the propagation delay time of the inverter. From this simple example it is possible to see that logic devices can be built into circuits to make other devices.

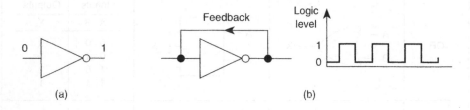

(a)   (b)

Fig. 9.5(a) and (b)

THE SET and RESET (SR) BI-STABLE or LATCH

Logic gates are built into integrated circuits to make other logic devices such as bistables or latches. They are referred to as bistables because they have two stable states. In the case of the SR these states are SET and RESET. In terms of positive logic the SET state sets the latch output at logic 1, and the RESET state sets the output at logic 0.

The SR latch shown in Fig. 9.6 is built from two NOR gates linked in a circuit as shown in Figs 9.6(a) and 9.6(b). The inputs and outputs are shown and the operation may be followed by using the truth table in Fig. 9.4. The circuit symbol for the SR latch is shown in Fig. 9.6(c). The line over the Q means 'not Q', or the opposite of Q. The output Q will always be 0 or 1 provided that the design is made so that S and R cannot be 1 simultaneously. Figure 9.7 shows an SR latch in an automatic headlamp circuit. This is included to show a motor vehicle application but I would not expect readers to spend too much time trying to fathom out how the circuit operates at this stage.

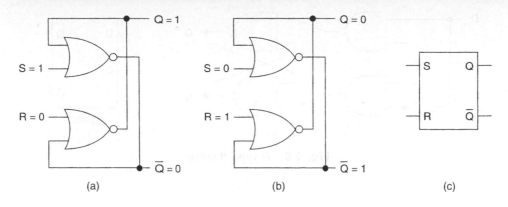

(a) (b) (c)

Fig. 9.6

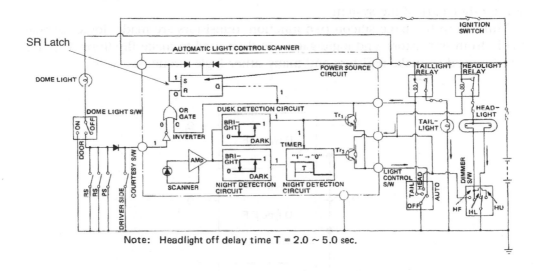

Note: Headlight off delay time T = 2.0 ~ 5.0 sec.

Fig. 9.7   Automatic headlamp circuit (Toyota)

THE D TYPE LATCH

Figure 9.8 shows the circuit for one form of D type latch. The D type latch (flip-flop) may be used as a basic memory element for the short term storage of a binary digit (0 or 1) applied to the data input D. The memory effect is achieved because whatever data (0 or 1) is at input D is transferred to the output Q by the clock pulse. The data thus transferred to Q is then held there until the next clock pulse.

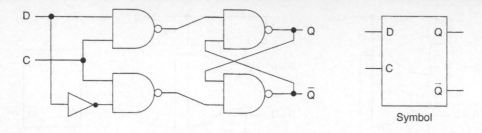

Fig. 9.8 D type flip-flop

REGISTERS

When several D type latches are joined up, as shown in Fig.9.9, a register is formed. The register shown has the binary word 1010 applied to its D inputs, with the most significant bit (MSB) at D_3. When the clock pulse arrives the data (D_1 to D_3) will be transferred to the outputs (Q_0 to Q_3). A register can be used as a temporary memory to hold data until a clock pulse arrives to pass the data to some other part of the system.

Thus, from the basic silicon p–n junction, transistors are made, logic gates are made from transistors, and logic gates are then used to make flip-flops. Flip-flops can be used to make registers and many other logic circuits as well.

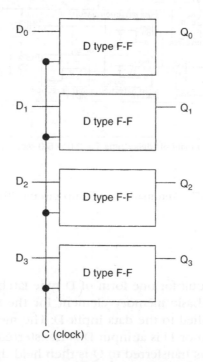

Fig. 9.9 D type flip-flops used as 4-bit register

Memories

READ ONLY MEMORY (ROM)

A read only memory (ROM) is an integrated circuit which consists of an array of transistors and diodes connected together to store an array of binary data (0s and 1s). When the binary data has been stored in the ROM it can be read out when required, but it cannot be changed under normal operating conditions. Figure 9.10 gives an indication of the way a ROM works. The ROM circuit truth table shows that for a given binary input there is a specific output. This output will only occur when the correct input pattern is received.

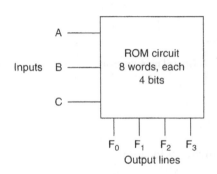

Inputs			Outputs			
A	B	C	F_0	F_1	F_2	F_3
0	0	0	1	0	1	0
0	0	1	1	0	1	0
0	1	0	0	1	1	1
0	1	1			·	
1	0	0			·	
1	0	1			·	
1	1	0			·	
1	1	1	0	1	0	1

Data stored in ROM circuit

Part of a ROM truth table

Fig. 9.10 Basics of a ROM

The microprocessor I/C part of an ECU will probably have three types of memory on the chip. These are likely to be a ROM, a RAM and an EEPROM. The largest of these, in terms of bytes, is the ROM. The ROM holds the program and fixed data, and a self-check program which can be operated to check many parts of the system automatically. The EEPROM can be re-programmed to permit non-volatile storage of variable data, e.g. different settings for various applications. The RAM acts as a temporary store for the results of calculations, comparisons, etc. that are made during the course of operation.

Figure 9.11(a) shows the basic principle of a memory cell that can be used in a RAM. The cell shown is a dynamic memory (DRAM); it consists of a transistor switch connected to a capacitor. The transistor allows the capacitor to be charged or discharged when the data is written into the cell, and also allows the presence or absence of an electrical charge (logic 0 or 1) to be detected.

FAULT CODE MEMORIES

Additional RAM is contained in most vehicle ECUs so that any malfunction in a system can be recorded while the vehicle is in operation. The information about

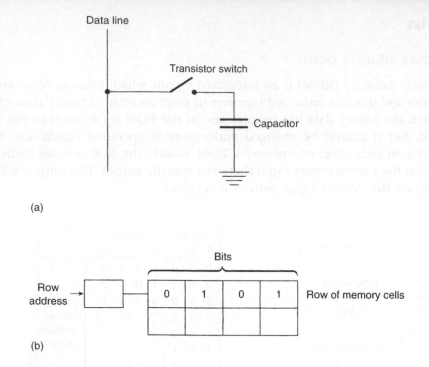

Fig. 9.11 (a) A DRAM cell; (b) a 4-bit memory location

faults is stored in binary form and is output in a variety of ways, some of which are covered in Chapter 6.

Vehicle ECUs are normally energised by a main relay which is controlled by the ignition switch, or the equivalent on a diesel vehicle. When the ignition is switched off, or the battery is disconnected, the ECU is de-energised and the contents of RAM are lost. If a fault code had been stored this would also be lost. Fault codes on these systems can only be read when the ignition is switched on and the system is operating, or when the system is stimulated externally to simulate operation. There are obvious advantages to be gained from systems that are able to retain fault codes for later analysis, especially so with intermittent faults.

Figure 9.12 shows an arrangement that partly overcomes the problem. The fault code memory is kept alive (KAM), keep alive memory, by being fed directly from the battery, via a fuse. When the ignition is switched off, the main part of the ECU is de-energised while the KAM remains active, until the battery is disconnected.

EEPROM (E²PROM)

The electrically erasable programmable read only memory (EEPROM) is a circuit that can be programmed by the ECU processor while in operation. This means that fault codes can be written to it during operation and they remain available to be read out at any time until the EEPROM is cleared.

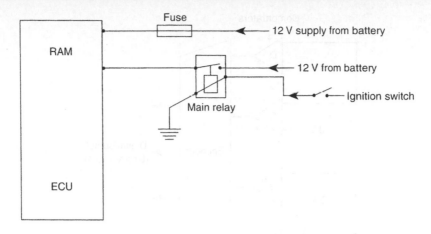

Fig. 9.12 Semi-permanent RAM (KAM)

Clearing of the memory is achieved by subjecting the EEPROM chip to an electric pulse (probably 20 V which is generated by a special circuit inside the controller). (*Note*. Not just any electrical pulse but *only* the one that is applied by using the manufacturer's instructions for clearing the fault code memory.)

Once again, this highlights the need to know specific detail about the product being worked on.

A/D conversion

The outline of an ECU in Chapter 2 shows an analogue to digital converter which takes voltage readings from various sensors on the vehicle and converts them to digital (binary 0s and 1s) which are used by the processor. Figure 9.13 shows this interface.

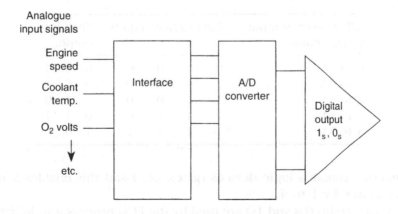

Fig. 9.13 ECU input interface

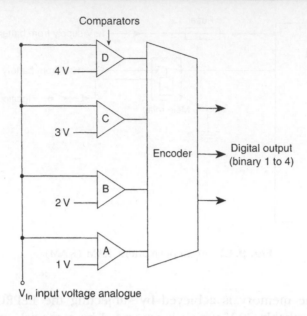

Fig. 9.14 Flash type analogue to digital converter

Conversion from an analogue voltage to a digital code (word) can be done in a number of ways. Figure 9.14 shows one type of A/D converter known as a 'flash' converter.

The flash converter shown in Figure 9.14 consists of four comparators and an encoder circuit which takes the comparator outputs and converts them into a binary code.

An electronic comparator is a circuit which continuously compares two signals. One of the inputs, at each comparator, is a reference voltage. When the input voltage matches the reference voltage the comparator outputs a logic 1. The reference voltages shown here are 1 V up to 4 V.

The following table shows the input/output performance of the converter.

A/D converter input voltage range	Comparator outputs				Encoder output		
	A	B	C	D			
0 to 1 V	0	0	0	0	0	0	0
1 to 2 V	1	0	0	0	0	0	1
2 to 3 V	1	1	0	0	0	1	0
3 to 4 V	1	1	1	0	0	1	1
4 to 5 V	1	1	1	1	1	0	0

The encoder contains logic devices (gates, etc.) and this enables it to output the binary codes for 1 to 4.

These binary codes (0s and 1s) are used by the ECU processor to initiate certain actions. They are moved around the ECU by buses (wires) and they consist of very low current electrical pulses. When a binary output command is generated it is normally required to be in analogue form. This requires a digital to analogue converter at the ECU output interface.

Digital to analogue conversion

Figure 9.15 shows the basic principle of a digital to analogue converter. The power sources of 8 V, 4 V, 2 V and 1 V are represented by the small circles. When a binary code is presented at the input, with the most significant bit (MSB) at the 8 V end, and the LSB at the 1 V end, the switches are operated electronically. In the diagram binary 1100 (12) is placed at the inputs. This means that the two inputs of 1 will switch their respective voltages of 8 V and 4 V to the output lines, and the electronic summing circuit will add them together to give a 12 V output.

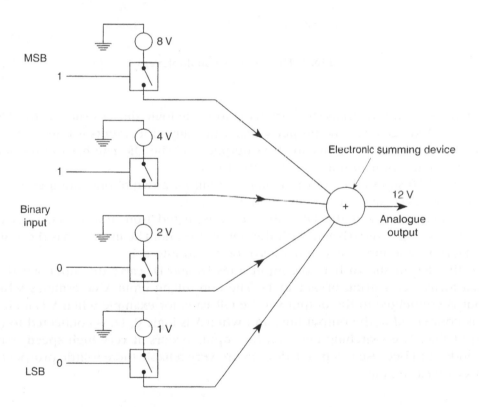

Fig. 9.15 Basic principle of a digital to analogue converter

The multiplexer

The multiplexer is an electronic device that allows a number of data inputs to be connected to a single output, one at a time, as dictated by some controlling inputs. The name multiplexer means 'many into one'.

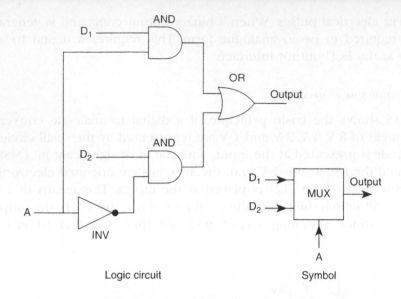

Fig. 9.16   Two to 1 multiplexer

I have shown how transistors are used to make logic devices such as the AND gate. The AND gate has the property that when any of its inputs is 0 its output is 0, and when all of its inputs are 1 its output is 1. The OR gate has the property that when any of its inputs is 1 its output is 1.

Figure 9.16 shows the electronic logic circuit for a two to one multiplexer that uses some AND gates, an OR gate and an inverter.

The AND gates and the OR gate are constructed from transistors, as is the inverter, which is the triangle with the dot on the end. The inverter has the ability to invert a logic input of 1 into a zero, or a zero into a 1.

In the design shown here the input lines D_1 and D_2 carry the input signals in serial form, i.e. a string of 0s and 1s. The controlling input A determines which input is connected to the output via the OR gate, for example when A is logic 1, D_1 is connected to the output line, and when A is logic 0, D_2 is connected to the output line. The switching between the inputs occurs at very high speed (nano seconds) and because electronic devices are very reliable there is little prospect of devices wearing out.

VEHICLE APPLICATIONS OF THE MULTIPLEXER

Instrumentation

The instrument display is the way in which the driver is informed of various factors relating to the vehicle's performance at a given time. The most common ones are vehicle speed (speedometer), fuel tank contents (fuel gauge), engine oil pressure (oil pressure gauge) and engine coolant temperature (temperature gauge). Each of these gauges is dependent on a sensor input, and for some years now a single processing unit (ECU) has been used to process sensor signals and present them as a display at the instrument panel.

In order to keep the ECU as simple as possible a multiplexer is used so that one sensor variable at a time is processed into an instrument panel display. If we imagine that our data inputs D_1 and D_2 are digital inputs from the coolant temperature and vehicle speed sensors respectively and that the processor can deal with only one input at a time, then it will be seen that the control input will put one or the other on to the multiplexer output. The MUX works at very high speed and can switch between the inputs for as long as is required. This means that a single processor can be used for all instruments, instead of a separate one for each.

Figure 9.17 shows the relation between the various parts of an electronic instrument display. The demultiplexer (demux) takes the single output from the processor (ECU) and directs it to the appropriate instrument. It works in much the same way as the multiplexer except that it takes one data input, from the ECU, and directs it to the required gauge. This means that the ECU is processing sensor inputs in a sequence, and selecting a gauge to send the output to, in the same sequence, which means that the gauge inputs are interrupted. It has to be remembered that the multiplexer and the ECU may work at very high speeds and, depending on the type of instrument used for the gauge display, this means that the intermittent nature of the gauge input does not produce discernible fluctuation on the gauges. For example, a fuel contents gauge can operate by sampling the sensor input every few seconds, whereas the speedometer will need to

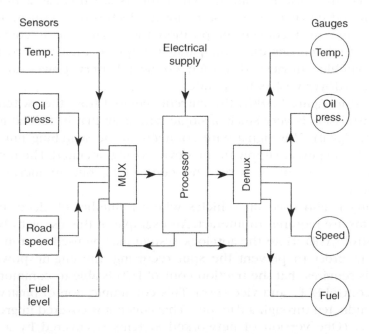

Fig. 9.17 An instrument system using a multiplexer

sample the sensor input several times a second. Such timing operations are easily performed by the timer section of the ECU.

Another vehicle use of the multiplex concept

As the number of electronically controlled systems on vehicles has increased so has the number of circuits, and this has led to a significant increase in the total length of wire used on the average vehicle. Coupled with the increased use of wire is an increase in the complexity of circuits.

In order to overcome these difficulties, systems have been developed in which a power line (bus) carries the electricity supply around the vehicle and output devices, such as motors, solenoids, etc. are connected to the power line, as and when required, by coded data signals which are transmitted via a signal line (data bus). A data bus is a wire or set of wires along which are transmitted data bits, 0s and 1s, for control purposes. When a code, such as 1011, is sent, one bit at a time along a single wire bus, it is known as serial data transmission.

Figure 9.18(a) shows a simplified form of conventional wiring where each electrical device on the vehicle has a separate circuit. Such a system requires many wires and connectors, and often the wires have to carry heavy current, e.g 25 amperes, so that the length and diameter of the cables can be quite difficult to accommodate in the restricted space on a vehicle. The use of numerous cable connectors can also be a source of problems, and it is natural that designers have sought ways of reducing the amount of wire and connectors used on a vehicle.

Multiplexed wiring has the capacity to overcome the difficulties and Figure 9.18(b) shows a simplified form of multiplexed wiring. Examination of this figure will show that the various control switch inputs are used as a binary coded parallel input and this is output in serial form which is then transmitted along a single wire (bus). At a convenient position the serial code is converted to a parallel output (in the demux). This parallel output is then used to trigger the electronic switching devices that then connect battery power to whichever application the driver wishes to operate.

As indicated in Figure 9.18(b) the current required from the switches and for the data transmission is very small; a single wire is all that is needed for this low current (data signal). This signal wire connects to the switching unit which, in turn, connects the power supply to the device being operated. The net result is a reduction in the amount of wires and connectors, but an increased use of electronic devices.

Multiplexing is also used on vehicles where a number of electronically controlled systems are required to interact. An example of this use is to be found in traction control (ASR). Here the anti-lock system may be used to stop a spinning wheel and, in order to prevent the spin recurring, the engine power may be reduced. This requires that the traction control ECU is able to communicate with the engine control ECU, and vice versa. This communication of binary codes (0s and 1s) is conducted through a data bus. This concept is covered in greater depth in Chapter 7. (One version of networked systems is covered by a procedure which has become known as CAN [Controller Area network].)

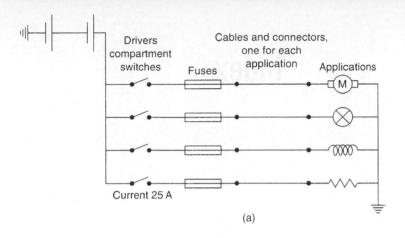

(a)

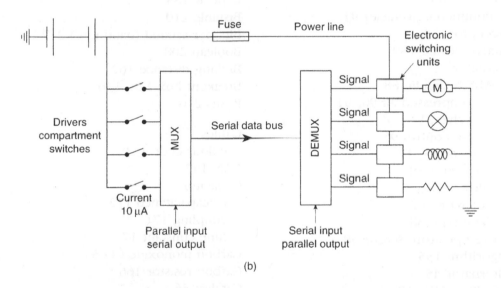

(b)

Fig. 9.18 (a) Conventional wiring; **(b)** Multiplex wiring

One of the concerns that repair technicians have about networked systems is that several systems may share signals from a single sensor which is electronically linked to the circuits. This may mean that the sensor can only be assessed for performance by means of the diagnostic system.

Index